Life in the Cold

LIFE IN THE COLD

An Introduction to Winter Ecology

PETER J. MARCHAND

with illustrations by Libby Walker

University Press of New England
HANOVER AND LONDON, 1987

University Press of New England

Brandeis University

Brown University

Clark University

University of Connecticut

Dartmouth College

University of New Hampshire

University of Rhode Island

Tufts University

University of Vermont

Library of Congress Cataloging-in-Publication Data

Marchand, Peter J.
　Life in the cold.
　Bibliography: p.
　Includes index.
　1. Cold adaptation.　2. Winter.　3. Ecology.
I. Title.
QH543.2.M37　1987　　574.5'42　　87–8222
ISBN 0–87451–416–9
ISBN 0–87451–417–7 (pbk.)

TO JANET

who has endured so unselfishly
all these winters, never quite sharing
my enthusiasm for the cold and snow.

TO GREG AND DANIELLE

who have, no doubt, gone without
for their father's obsession with science,
never quite understanding what it is
that I do for a living.

AND TO MY STUDENTS

of winter ecology past and future,
who always ask the right questions
and whose enthusiasm keeps me going
on the really cold days.

Take winter as you find him—and he turns
Out to be a thoroughly honest fellow with
No nonsense in him

And tolerating none in you
Which is a great comfort in the long run . . .

From Tass's winter cabin
Author unknown

CONTENTS

PREFACE

It has always struck me as odd that traditional biology and ecology programs give so little attention to the special problems of plants and animals wintering in the North, especially considering that so much of our landscape lies under snow and ice cover for so much of the year. I suppose that there are many reasons for this. Traditional academic calendars, for one thing, do not allow much time to be spent in the field during the winter months. Even when time is available, the logistical problems of getting people and sampling equipment to study sites and of keeping that equipment (and sometimes the people) operating in the snow at subfreezing temperatures, are often rather discouraging. To me, however, this has been more a justification for the effort than a deterrent. And it's not that winter biology is unknown. To the contrary, there are a great many noteworthy research efforts recorded in the literature. The problem, though, is that the individual pieces are rarely integrated to make up the whole; and that, of course, is what ecology as a discipline is all about. So I have taken a small step here toward that end by piecing together a more comprehensive picture of organisms, both plant and animal, interacting with their environment and with each other in ways that are often unique to the season. The feelings of humble enlightenment as more and more pieces come together have made my winters much more enjoyable. This insight is the best that I can hope to pass on through the writing of *Life in the Cold.*

As in any discourse of a scientific or philosophical nature, it seems appropriate at the outset to define the bounds of the subject, if only to focus initial arguments. "Winter," to one degree or another, comes to every land beyond the tropics as the earth on its tilted axis alternately dips the Northern and Southern hemispheres toward and then away from the sun in making its annual rounds through space. In some places, the rhythm of life changes little

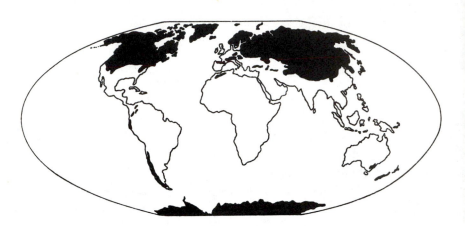

World snowcover distribution. Shaded areas are locations having snow on the ground for more than two months of the year, up to and including permanent snow and ice. (Data from G. T. Trewartha, An Introduction to Climate, *4th ed. [New York: McGraw-Hill, 1968].)*

through it all and one has to look carefully for the subtle signs that say this season is different from the rest. Elsewhere winter is the principal architect of life-form and habit. At higher elevations and higher latitudes, winter is a season of chilling energy deficits that demand the most conservative of physical and physiological adaptations in plants and animals. This is the winter I have in mind in writing this book. For the most part, then, I shall confine my discussions to those geographic areas where winter temperatures are low enough to elicit consistent and predictable *adaptive* responses from terrestrial organisms, and where snowcover is of sufficient duration to be considered an integral part of the environment (see map) and is important enough to make a difference in the overwintering success of the plant or animal. I shall let those two conditions define "winter" and prescribe the geographic coverage of this book.

This is by no means a conclusive treatise on the subject. Rather it is something of a progress report. I have been a student of winter ecology for nearly 15 years now, and I am just beginning to understand where the crucial matters lie. *Life in the Cold,* then, is as much a statement of what we don't know as what we do know. Throughout the book I have attempted, in a manner sometimes uncharacteristic of the sciences, to convey a sense of the freshness and personal excitement I feel toward the subject. In order to accomplish this, I have tried to reduce the formidable technicality of some material and remove, as much as possible without sacrificing accuracy, the veil of scientific jargon that sometimes inhibits, rather than helps, the learning process. In some cases I have stayed with familiar terms, even if they are less desirable from a scientific point of view, for example, where it seemed more convenient, retaining the terms "warm-blooded" and "cold-blooded" for their respective equivalents "homeotherm" and "poikilotherm." I have, however, used the international system of units throughout the text, hoping that it will not inconvenience my lay readers in countries still using English units. Despite my best attempts at simplification, it will be awkwardly clear in places that my primary commitment is still to the science rather than the prose.

Some may find *Life in the Cold* wanting in coverage of specific organisms. I have found it desirable, though, both in my writing and in my research, to limit my initial attention to problems of what I consider to be "winter-active" organisms—those that must face the rigors of winter on a day-to-day basis. I include plants as winter-active for reasons that will soon be obvious. In fact, early in my teaching I discovered that it is in the area of plants and the winter environment that the greatest misunderstanding (or lack of understanding) lies, and so I have treated certain aspects of plant ecology in considerable detail. On the other hand, because much more has been written elsewhere on the subject of animals in winter, I have confined my writing here to a few basic problems related mostly to the energetics of winter-active animals. If I succeed here in dispelling a few misconceptions and raising a few more unanswered questions that I think are important (or at least interesting), then I will be content for the moment. Once I have you thinking critically about winter ecology, then whatever deficiencies surface in my writing will no longer matter.

Though I take full responsibility for the content of *Life in the Cold,* I didn't get this far without a great deal of help. As a mentor and colleague of mine in the sciences often reminded me, "We all stand on the shoulders of the giants who walked before us." There have been many giants in my life. More than a few have lent a shoulder in the preparation of this book, some indirectly through their own contributions to the literature and others more directly through their enduring friendship and support, and through patient editorial and clerical assistance. Thanks to all, especially to Gerard Courtin of Laurentian University, whose critical review of this work went far beyond the call of duty. My debt is great.

Now it is my sincerest hope that the reading that lies ahead of you will enrich your own experience in the winter environment as much as the learning has enriched mine.

Johnson, Vermont   P.M.
February 1987

UNIT CONVERSION FACTORS

To convert from	to	multiply by
kilometers	miles (statute)	.621
kilometers	feet	3280.84
meters	feet	3.281
meters	inches	39.37
centimeters	inches	2.54
kilograms	pounds (avdp.)	2.205
grams	ounces (avdp.)	.035
liters	cubic feet	.035
liters	gallons (U.S. liquid)	.264
liters	quarts	1.057
milliliters	cubic inches	.061
milliliters	ounces (U.S. liquid)	.034
joules	calories (gm)	.239
joules	watt-seconds*	1
watt-hours	calories (gm)	860.421

*Note that the watt is a rate of work or power—specifically, that power that gives rise to the production of energy at a rate of 1 joule per second.

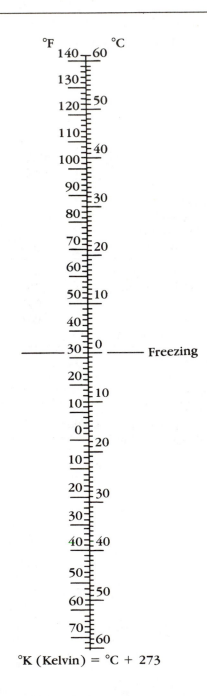

°F   °C

140 — 60

130

120 — 50

110

100 — 40

90

80 — 30

70 — 20

60

50 — 10

40

30 — 0 ——— Freezing

20

10 — 10

0

10 — 20

20 — 30

30

40 — 40

50

60 — 50

70

— 60

°K (Kelvin) = °C + 273

Life in the Cold

1 WINTER PATHS: OPTIONS FOR OVERWINTERING SUCCESS

In the lair under the uprooted old spruce, a heart beats faintly but steadily as the season marches on. All is quiet now, yet there is so much life in that still, dark hulk. Soon, barely waking, she will give birth to two or three cubs that will suckle in their mother's warmth until they are as full of energy as the sunlight bouncing off the late-winter snow. It is just one of many landmark events passing unnoticed in a natural world that has paused from its usual business to carry out a few of its more important acts in relative privacy.

Maybe the hardest notion to accept about winter is that it is so alive. Beneath the bark of the leafless tree, under the frozen moss, in all the little crevices of winter, there is life! For all the apparent calm of the winter landscape, a continuous play of activity is enacted daily behind the great white curtain. A masked shrew hastily nuzzles its way through the leaf litter, knocking down the fragile ice crystals that hang from the ceiling of the protective snowpack, almost frantically searching out food to fuel its metabolic furnace. Spiders, too, remain active under the snow, and with a little luck the shrew finds plenty to eat. But others have the same thing on their minds and there's nothing secure about the life of a shrew under the snowcover. With no more than a flick of a shadow to forewarn its prey, an ermine, diving down through the snow as adeptly as a cormorant at sea, comes up with the shrew and returns to line its own nest with the fur of its quarry (the ermine is warmed twice for its hunting skill).

The ermine, in its yellow-white winter pelt, is as natural to this scene as the snow itself. Almost from childhood we are taught that here is a creature truly at ease in the northern winter. But why? Its body is really all wrong for cold climates. Long and skinny, this animal is poorly adapted for conserving heat. While its short fur may keep ladies of fashion warm during their ever-so-brief encounters with winter, it hardly seems adequate insulation for subzero nights on the hunt. On the other hand the ermine *is* white, and we have the feeling that white in winter is good. But again, why? Protective coloration perhaps; but what does such a swift and voracious carnivore that spends much of its time in winter hunting the dark recesses under the snowpack need of protective coloration? Could it be that white confers some other not-so-obvious advantage?

As an object of winter ecology, perhaps we could choose no better symbol than the ermine, for it seems to represent all that is right *and* wrong in a winter-active animal. It reminds us that, within limits dictated by certain physical realities, there are many compromises in nature—that the forces shaping life in snow country often evoke different strategies, none of which solve all problems, but each of which serves a particular organism in terms of its own role in the scheme of things. Even among plants we find many alternative, often starkly contrasting solutions to the problems of winter. As we look closely, there too we find some common threads. Certain inventions work, others don't, and it is by trial and error that they are sorted out.

Through the endless process of natural selection, three basic strategies for surviving winter's rigors have evolved: migration, hibernation, and resistance. There are variations in theme, of course, but in one form or another these are the options. What they amount to in an evolutionary context is a choice between avoidance and confrontation. In a broader sense, though, each of these options confronts the season in its own way; even the "choice" of leaving a frozen landscape in favor of more southern latitudes is, in reality, a desperate stand against winter.

MIGRATION

At first glance, migration would appear a safe alternative to wintering in the North—an escape clause, as it were, in the contract between an organism and a demanding environment. But migration is a very precarious business. To begin with, the energetic cost of traveling long distances is extremely high. In order to fuel a 1000-km flight, a bird preparing for a migratory trip must accumulate so much reserve energy that it may carry at departure up to 50% of its total body weight as fat.[1] Adding weight for flight has its practical limitations, however, and even this preparation may not be enough.

The energy reserves of migrant birds flying over water sometimes become exhausted before they reach their destination.

Various other obstacles also confront the migrating bird. For some waterfowl there is the additional stress imposed by humans through unrelenting hunting along the migratory route. A Canada Goose leaving James Bay in early September, headed down the Atlantic or Mississippi Flyway, may face up to four months of shooting as the legal goose season opens progressively later southward along the route. When the survivors of this journey arrive at their destination, they face still other uncertainties. A new assemblage of parasites, diseases, and predators, along with changing food availability and difficulty in forming efficient new search habits, may take an additional toll.

Migration, thus, is not an easy out; but for many bird species there is no alternative. Physical or behavioral adaptations to particular feeding strategies alone may dictate fall flight. The herons, for example, with their stilt-legged manner of fishing for a living in shallow water, have no way of coping with even a thin, temporary cover of ice. They have, in effect, become too specialized. The flycatchers as well, once their insect prey have metamorphosed and become sedentary for the winter, must move southward to find food on the wing. And so, too, must the soaring birds of prey. When the northern landscape becomes uniformly white under snow and ice cover, there is no longer any differential heating of the earth's surface to generate uplifting air currents so that these predators may ride the thermals with minimal energy expenditure while they hunt.

Migration, then, is a necessary risk for some organisms. For others it is not a viable option at all, if only because the cost of transportation may be too great. In energetic terms, it is far more expensive to travel overland than to fly. It has been estimated that a mammal would expend approximately 10 times more energy moving a given distance by running than would a bird of equal weight flying that same distance.[2] This may be one reason why mammals in the North generally do not migrate to avoid winter. Interesting exceptions to this generalization include: caribou, which often move several hundred kilometers between their summer breeding grounds on the tundra and their winter shelter in the boreal forest; species of bats that migrate considerable distances to reach their winter refuge (using a more efficient means of getting there, of course); and such marine mammals as the gray whale, which may cover the entire Pacific coastline of North America at an energetic savings even greater than that of flying.

HIBERNATION

The alternative to migration as a means of avoiding many of winter's stresses is hibernation, though hibernation is not without its risks, too. The hibernating mammal possesses an unusual ability as a "warm-blooded" animal or "homeotherm" (an organism that maintains body temperature independently of surrounding temperature) that can enter a state of much reduced metabolic activity, allowing its body temperature to track surrounding temperature in much the same way as, say, a turtle or other "cold-blooded" animal. When air or ground temperature falls, so too does body temperature. The hibernator, though, generally cannot allow its body temperature to drop below freezing. When so threatened, it must arouse from its sleep and restore its body temperature all the way to normal before reentering the hibernation state. This is very costly, in energetic terms, because the animal must metabolize a considerable amount of its fat reserves in order to rewarm. There are some exceptions— hibernators such as the European hedgehog can maintain a body temperature just above freezing instead of rewarming completely— but in general, it is critical to the success of the hibernating animal that it find a refuge where the temperature is not likely to drop below 0° C. As one proceeds northward, the availability of such safe hibernacula becomes much scarcer, particularly in permafrost regions. It is not surprising, then, that hibernation as an overwintering strategy is relatively uncommon among mammals in the Far North. If one plots the geographical distribution of true hibernators, it becomes apparent that this is largely a midlatitude winter adaptation (fig. 1).

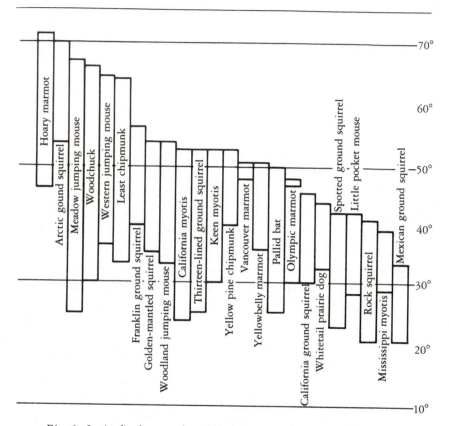

Fig. 1. Latitudinal range of true North American hibernators. Because of the relative scarcity of "safe" hibernacula in the Far North, hibernation is more common among midlatitude mammals.

For reptiles and amphibians of snow country the options are much more limited. These "cold-blooded" animals or "poikilotherms" have little control over their body temperature, and when the days and nights get cold their metabolism slows to the point of near helplessness. For the majority of these organisms, hibernation is the only means of surviving the winter season. As the days of fall become shorter, reptiles and amphibians seek out safe winter hiding places—sometimes under decaying logs, in deep protected rock crevices, or in the burrows of other animals—wherever they can escape subfreezing temperatures. Snakes sometimes congregate in groups of mixed species: rattlesnakes, copperheads, and northern black racers are frequently found together in rock dens. In unusual cases, countless thousands of a single species will congregate in deep caverns (such as those found in the limestone country of southern Manitoba), some individuals traveling 16 km to reach the same hibernaculum year after year.[3]

Frogs and toads may also seek underground burrows for hibernation, the latter frequently digging themselves into soft soil with their hind feet to a depth of nearly 1 m. Some have been known to use caves, too. Others take advantage of stream banks and seepage areas where temperatures are not likely to drop below freezing, or burrow into the soft mud of pond bottoms, sharing this winter refuge with salamanders and turtles. Occasionally reptiles and amphibians overwinter with little more protection than a cover of moss or leaf litter, and there they have been observed to survive temperatures of a few degrees below freezing. In fact, some species of land frogs tolerate ice formation in the intercellular spaces of their body tissues, while protecting the living cells with glycerol.[4] But exposure to subfreezing conditions puts most in a precarious position. Any disturbance while in a "supercooled" state, where body tissues are dangerously chilled without ice formation below their actual freezing temperature, may result in immediate death. As was the case with hibernating mammals, the overwintering success of reptiles and amphibians is best assured by avoidance of subfreezing temperatures.

RESISTANCE

To many organisms winter means staying and enduring, facing the rigors of the season and resisting its stresses. Consider the sedentary tree or the leafless shrub with ice-covered buds poking above the snow. Are these totally dormant in winter, insensitive to the chill winds of January? Do they pass the season in frozen animation with their biological clocks set to resume life upon some proper cue in the spring? Or do they remain sensitive to their environment, able to respond to favorable circumstances during the winter? And consider the caterpillar overwintering in the frozen woodpile. Its emergence in the spring tells us it has remained very much alive. Yet these and so many other insects in their larval and pupal stages must withstand temperatures that would kill most living cells. Because these organisms produce no appreciable heat, they must depend upon complex biochemical mechanisms that enhance their ability to survive freezing temperatures. The production of glycerol by some insects, for example, inhibits ice formation in their body tissues, permitting them to cool without freezing, sometimes to temperatures as low as $-50°$ C. In others the production of special ice-nucleating proteins promotes extracellular ice formation and reduces the risk of flash freezing, which is always a danger with supercooling. We think of these organisms as hibernating to avoid winter, but in reality theirs is a complex strategy for *resisting* severe cold stress.

For winter-active birds and mammals, resistance often involves coping with snow and ice cover in the conduct of their daily affairs. Some animals are superbly adapted to the physical constraints imposed by snow, while others are substantially impaired by it. Lemmings develop a digging claw to aid them in foraging under the snow, and spruce grouse grow extra scales on their feet to enable a better grip while feeding on icy branches. The caribou, with feet seemingly way out of proportion to its body size, floats over the snow as if on snowshoes (fig. 2*a*). The lynx and snowshoe hare also benefit from the efficiencies of a high foot-surface-to-body-weight ratio. The moose, on the other hand, confronts deep snow differently, with a stature and musculature that allow it to lift its feet almost shoulder high, thus minimizing the amount of energy expended in wading through the snow. In marked contrast, the white tail deer suffers immeasurably in deep snow, unable to stay on the surface and unable to lift its legs clear (fig. 2*b*). The track of the whitetail shows characteristic foot drag in as little as 15 cm of snow. This inability to deal with deep snow explains in large part the

a   *b*

*Fig. 2. Coping with deep snow. Caribou and their domesticated counterparts,
the reindeer (a), are known as "floaters"—their oversized feet enable them to move
easily over snow-covered ground. In contrast, the mobility of the whitetail deer
(b) is greatly hampered by deep snow, forcing them often to congregate in large
"yards," usually under coniferous forest cover. Their collective trampling of
the thinner snowcover here may allow more freedom of movement, but food resources
soon become scarce in the vicinity of the yard and starvation is often their fate.
(Photo credits: 2a by Peter Marchand; 2b by John Hall.)*

yarding and feeding behavior of deer in northern areas during
winter.

In all, there exist a number of interesting adaptations in the animal world for coping with snow. But for many organisms the more
serious problems of winter are associated instead with extreme cold,
and for their survival a dependable annual snowcover is of utmost
importance. Snow is both the reflective cover that turns away the
sun's warming radiation and the thermal blanket under which much

biological activity takes place during winter. While it is the bane of those plants that must bear its weight and those animals that must scratch through it for sustenance, it is the salvation of many other plants and animals that depend upon it for protection from the cold. Snow separates two worlds, ours and the "subnivean," that world beneath the snowcover. So important is it to both environments that William O. Pruitt, Jr., the well-known biologist from the University of Manitoba, considers the initial accumulation of 20 cm of snow to mark the true beginning of winter—the "hiemal threshold," as he terms it. It seems appropriate, then, that this present treatment of winter ecology should also begin with the subject of snow, considering first the physical charcteristics of the snowpack and then probing the true nature of the environments above and beneath the snow surface.

2 THE CHANGING SNOWPACK

Snowflakes, for all their elaborate detail, are destined for destruction almost from the time they form. The molecules that make up the delicate needles and plates of multifaceted crystals are too active to remain long in such intricate formation. There are plenty of outside forces, too, that act upon the snowflake to hasten its destruction. Tumbled in the wind, fractured and compacted, vaporized and recondensed, and melted and frozen again, snow on the ground is constantly undergoing change or metamorphism. This not only alters the individual ice crystal, but also changes the internal structure of the snowpack.

Collectively, snow crystals are affected by at least three distinct processes in their metamorphism, and each process is influenced by time, internal snowpack characteristics, and external weather conditions. These processes are described here in the sequence in which they might occur after initial deposition, though it is possible for the three processes to occur in any order—or even simultaneously—within different parts of the snowpack.

DESTRUCTIVE METAMORPHISM

The initial deterioration of the snowflake and resultant formation of more or less rounded ice grains in the snowpack is referred to as "destructive" or "equi-temperature" metamorphism. The term "destructive" is self-explanatory as it pertains to the fate of individual snow crystals within the snowpack. The qualifier "equi-temperature" serves to distinguish this process from the next, which requires an unequal temperature distribution within the snowpack. Destructive metamorphism involves a reordering of water molecules on the surface of each snowflake with a resulting loss of fine structure—a deterioration of the radiating arms of delicate needlelike crystals, for example. This reorganization in effect represents a redistribution of molecular energy within the ice crystal. The snowflake is thus transformed into a roughly spherical particle with a net reduction in surface area. The resulting ice grains may than coalesce with others, the smaller ones being absorbed into larger ones, until all are nearly

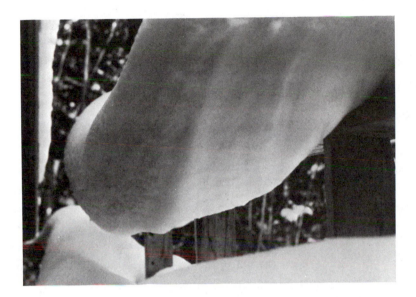

Fig. 3. Snow metamorphism. The increased mechanical strength of snow due to the bonding of individual ice grains—a process called "destructive metamorphism"—is evident in this picture of snow on a railing.

the same size. This process is influenced by anything that promotes closer contact of the ice crystals, such as wind packing or the weight of overlying snow, and is affected somewhat by temperature. Destructive metamorphism proceeds faster when air temperature is relatively warm than when air temperature is very low.

Throughout the process of destructive metamorphism snowpack air spaces are reduced in size as individual ice grains pack together and bond at their points of contact. Both the snowpack density and the mechanical strength of the snow increase substantially by this process, often within just a few hours after a snowfall (fig. 3). This is of some practical interest as it pertains to the use of snow shelters in the North. The traditional shelter of the Inuit winter camp, the igloo is constructed of wind-packed snow that owes its strength to the bonding that takes place during this metamorphic process. The effectiveness of the Quin-zhee, a snow house of Athapaskan origin used in boreal forest regions, is also related to the increase in strength accompanying destructive metamorphism when loose snow is piled and allowed a short while to "set" naturally (see box on "Snowpack Metamorphism and Throwaway Housing").

CONSTRUCTIVE METAMORPHISM

After the initial deposition and destructive metamorphism of snow on the ground, the snowpack begins to undergo changes in vertical structure through a process termed "constructive" or "temperature-gradient" metamorphism. Constructive metamorphism results from the migration of water vapor upward within the snowpack and depends on two snowpack properties: the existence of a temperature gradient between the bottom and top of the snowpack, and an interconnecting system of pore spaces within the snowpack.

A temperature gradient normally exists in the snowpack because of the low thermal conductivity of snow (a topic to be discussed in more detail later). The uppermost portion of the snowpack is affected by the cold air above, while the bottom of the snowpack is influenced by the ground, which provides a continuous source of heat throughout the winter. Because heat is conducted only very slowly through the snow, a pronounced temperature gradient is usually maintained between the colder top and warmer bottom of the snowpack. This temperature difference is important in that it affects the distribution of water vapor in the snowpack.

Within the pore spaces of the snowpack, water molecules are continually escaping from the ice surfaces until the internal atmosphere is crowded with vapor molecules and as many return to the ice surfaces by random bouncing as leave. Through this continual process of sublimation (the conversion of ice directly to water vapor) and condensation, the air spaces of the snowpack remain saturated—in other words, at 100% relative humidity. Because the amount of vapor in suspension at any given time increases with temperature, the warmer regions of the snowpack will have a higher concentration of water vapor molecules for a given volume of pore space than will

SNOWPACK METAMORPHISM AND THROWAWAY HOUSING

A Quin-zhee is a snow house of Athapaskan origin used in colder boreal forest regions of North America where snow is generally loose lying and of very low density. The shelter is constructed simply by shoveling snow into a large pile (*a*) and leaving it to set, without any additional packing, for an hour or two before hollowing out the inside living quarters (*b*). The

a

b

domed outside of the structure becomes self-supporting after a very short time because of a rapid increase in bonding strength between ice crystals, all due to the acceleration of destructive metamorphism caused by the mixing process and the weight of the snow itself. A person working alone can pile sufficient snow for a Quin-zhee in 30 minutes, and may begin digging out the interior only an hour or so later. A snowshoe is the only tool needed to construct a shelter that can warm to 25° above the outside air temperature. A very efficient means of obtaining shelter in deep fresh or unpacked snow, the Quin-zhee is easy to build and just as easy to leave behind, lessening the burdens of winter travel in the northern woods. (Photos by Peter Marchand.)

the colder portions. This means that, as long as there is a temperature gradient in the snowpack, there will also be a vapor concentration gradient. And because the molecules in suspension are continually bumping each other and trying to escape their confines, water vapor will tend to diffuse from areas of higher concentration to areas of lower concentration, which in this case means from warmer to colder regions of the snowpack. This migration via interconnected pores will go on even though the air spaces at the colder end of the gradient are saturated.

With the upward migration of water vapor through the snowpack, two things happen. First, condensation occurs in the upper, colder regions of the snowpack because the internal atmosphere is already at 100% relative humidity and the addition of more water vapor can only "supersaturate" the air. Condensation onto the outside of the ice grains in this region thus results in a growth of the grains. Secondly, water vapor migrating upward is continually replaced in the lower regions as sublimation from ice crystals maintains a saturated atmosphere there. This means that ice crystals at the bottom of the snowpack are continually diminishing in size.

Fig. 4. Depth hoar. As water vapor constantly migrates upward from warmer to colder regions of the snow, brittle, loosely arranged crystals known as "depth hoar" form at the base of the snowpack. Through this process of constructive metamorphism, more free space is created in the subnivean environment, facilitating the movement of small mammals, but at the same time contributing to avalanche danger in mountainous terrain. (Photo by Peter Marchand.)

The end result of constructive metamorphism is the formation, at the base of the snowpack, of very brittle, loosely arranged crystals referred to as "depth hoar" (fig. 4). The formation of depth hoar quite likely facilitates the movement of small mammals as they forage under the snow throughout the winter. Constructive metamorphism is also accompanied by a substantial loss of mechanical strength of the snowpack, and this has another consequence of considerable human concern. In mountainous regions the process of constructive metamorphism leads directly to an increase in avalanche danger. In times past when mountain troops were brought out to foot-pack threatening slopes, their job was, in effect, to reverse the constructive metamorphic process and promote destructive metamorphism instead.

MELT METAMORPHISM

The third process of snowpack alteration is "melt metamorphism," an almost self-explanatory change that comes about when any part of the snowpack is subjected to temperatures above freezing. As simple as it may appear, some of the finer points of this process are worth elaborating.

Surface melt may influence the internal temperature of the snowpack in a way that would seem out of proportion to the amount of melt taking place. This is due to the release of latent heat as meltwater percolating down into the snowpack encounters lower temperatures and refreezes. As already noted, under typical winter conditions the snowpack is warmest at the bottom and coldest at the top. However, as the air above the snowpack warms above freezing, the surface snow will become nearly the same temperature as the snowpack bottom, with a colder layer sandwiched between. This is related again to the fact that snow conducts heat relatively slowly, and changes occurring in the upper part of the snowpack may not affect lower portions right away. When surface meltwater percolates downward and refreezes in the colder stratum, it releases heat in the amount of 335 joules/g of water frozen—the latent heat of fusion (latent heat is the energy gained or lost with phase changes in water). The meltwater essentially acts as a heat pump, gaining 335 joules/g at the surface when it changes from solid to liquid and transferring this energy to lower layers of the snowpack when it freezes again. This has the effect of bringing the whole snowpack rapidly to equal temperature.

Melt metamorphism also occurs when rain falls on the snow surface or when fog forms over the snowpack. In the case of rainfall, the amount of heat energy transferred to the snowpack is related to the temperature of the rain itself and, of course, to the total quantity of rain entering the snowpack. These two measures give the total energy input of the rainwater, to which is added the latent heat released when this water freezes. In the case of fog, the amount of energy transferred to the snowpack is equal to the latent heat released as water condenses over the cold snow to form the fog droplet. The latent heat of condensation is quite high, amounting to approximately 2450 joules/g of water, depending on the exact air temperature. This means that, for every gram of water vapor condensed over the snow surface, enough heat energy is released to melt seven times as much ice (remember, the latent heat of fusion is only 335 joules/g). This helps explain why snow seems to disappear so much faster in fog than it does during a rain.

It should be apparent by now that the winter snowpack is a very dynamic entity, everchanging in its physical character. The processes described above continually alter the snowpack—constructive metamorphism breaking down old crusts, melt metamorphism forming new ones, fresh snowfalls starting the process over again with destructive metamorphism. As long as there is snow on the ground, these forces are at work. The result is a snowpack layered with a graphic history of winter's weather. Studying that record carefully may yield considerable insight into the problems experienced by organisms living on and beneath the snowpack.

THE INSULATIVE VALUE OF A SNOWCOVER

It has already been suggested that the winter survival of many organisms depends on the presence of an insulating cover of snow. The question arises, then, as to how much snow is enough. Is 20 cm, Pruitt's hiemal threshold, the critical depth? Or is more needed? The answer depends not upon depth alone, but also upon the degree of snowpack metamorphosis that has taken place. For among the various physical properties of snow that are altered by the processes of snowpack metamorphism the one that perhaps most affects life in the subnivean environment is changing snow density. Through its effects on thermal conductivity, density differences strongly influence the insulative value of a snowcover.

The thermal conductivity of snow is measured as the amount of heat passing through it over a period of time, for a given gradient in temperature between the top and bottom of the snowpack. The

Table 1. Thermal conductivities of various natural and man-made materials

Material	Density (g/cm)	Conductivity (W/cml°K)
Granite	2.70	.029
Ice (0° C)	.92	.021
Sandstone	2.25	.017
Concrete	2.85	.012
Glass	2.60	.008
Old snow	.40	.004
Wet peat	.90	.0033
Dry sand	1.55	.0021
Dry peat	.45	.0008
Sawdust	.19	.0004
Glass wool	.06	.0004
Fresh snow	.10	.0003
Still air	.001	.0002

Sources: R. Geiger, *The Climate Near the Ground* (Cambridge: Harvard University Press, 1965) and *Handbook of Chemistry and Physics* (Boca Raton, Fla.: CRC Press, Inc.).

lower the thermal conductivity of a material, the better its insulative value. A list of thermal conductivities for various natural and man-made materials is given in table 1. Notice that the thermal conductivity of snow is not only very low relative to many other natural materials, but it is also closely related to snowpack density. The higher the density of snow, the higher its thermal conductivity and, therefore, the poorer its insulative quality.

The insulating effect of a snowcover is most readily apparent in the reduction of daily temperature fluctuations underneath the snowpack. This effect can be assessed indirectly through calculation of a thermal index value that combines measurements of both depth and density (see box on "Estimating the Insulative Value of a Snowpack"). As thermal index values approach 200, the subnivean en-

ESTIMATING THE INSULATIVE VALUE OF A SNOWPACK

A quick and easy assessment of the insulative quality of a snowcover is possible through the use of a thermal index scale that integrates measurements of depth and density. Mathematically the index, I_T, is a simple summation of depth/density values for each layer of the snowpack, written as follows:

$$I_T = \sum_{i=1}^{n} (z/G)_i,$$

where z is the thickness (cm) and G is the density (g/cm^3) of each layer, i.* This expression simply says numerically what we have already seen about snow: its insulative value increases (hence, thermal index values will be higher) under snow layers of lower density or increasing thickness.

In practice, the thermal index is obtained by exposing a fresh vertical face of the snowpack, measuring the thickness of each discrete snow layer, and dividing the thickness by the density of the layer. The resultant values are then added for the whole profile. A snow sample may be collected with any kind of cylindrical corer (a small can opened at both ends works well) pushed either vertically or horizontally into the snowpack. The measured volume of snow thus obtained may then be weighed in a small plastic bag suspended on a spring balance to obtain snow density. With commercial coring devices such as the "Mt. Rose" or "Adirondack" snow samplers, the entire cylinder containing a vertical snow core is cradled on a balance, which is usually calibrated to convert snow density directly into water content. In the latter case, the thermal index can be approximated by sampling the snowpack with a single vertical core and dividing the total depth by a single density value. Thermal index values have proven useful in assessing the ecological significance of a snowcover, especially where it is marginal, or in early fall and late spring when snow conditions may be critically important to small mammals. It is a useful technique also for evaluating the impact of snow vehicles or similar disturbances on the quality of the subnivean environment.

*P. J. Marchand. "An Index for Evaluating the Temperature Stability of a Subnivean Environment," *Journal of Wildlife Management* 46 (1982): 518–20; and N. M. Kalliomaki, G. M. Courtin, and F. V. Clulow, "Thermal Index and Thermal Conductivity of Snow and Their Relationship to Winter Survival of the Meadow Vole, *Microtus pennsylvanicus*," *Proceedings, Eastern Snow Conference* 29 (1984): 153–64.

vironment is no longer influenced by short-term temperature changes above the snowpack (fig. 5). Further increases in thermal index value through the addition of more snow will make little difference to the temperature stability of the subnivean environment. It is interesting to note here that a thermal index value of 200 could result from a single snowfall of 20 cm depth where the density is .1 g/cm^3. This corresponds closely to Pruitt's hiemal threshold. Keep in mind, however, that the density of new fallen snow will increase over time through destructive metamorphism lessening its insulative value accordingly. On the other hand, once snow depth approaches 50 cm, changes in density up to .3 g/cm^3 become relatively unimportant in terms of their effect on subnivean temperature fluctuations because the greater depth compensates for the increased density. This is illustrated in figure 6, in which several snowpack temperature profiles, obtained with implanted thermistors, are plotted over a period of time during which considerable air temperature fluctuations occurred. Under 40 to 50 cm of snow, the temperature of the subnivean environment is almost constant.

For many small mammals, the presence of an adequate snowcover is critically important to their overwintering success. Restricted in their ability to add insulating fat and fur, they depend instead on the snowpack. Often their only means of maintaining a favorable heat balance in the face of decreasing air temperatures is to minimize

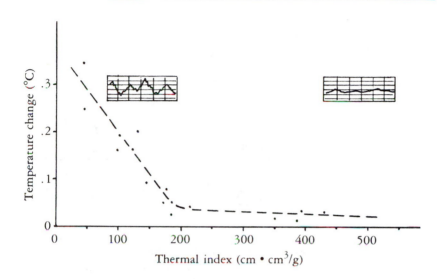

Fig. 5. *Temperature fluctuations under snow in relation to snowpack thermal index values. The y-axis is the ratio of 24-hour subnivean temperature change to 24-hour air temperature change. Insets at the top depict graphically the daily temperature changes under snow of greatly different thermal indices.*

body-air temperature differences by remaining under the snowpack, a strategy that will be discussed in some detail later. But the effect of snow on the energy relations of both plants and animals goes much further than its role as an insulative blanket, for snow greatly alters the partitioning of energy from sources external to the snowpack. To understand fully the influence of snow on winter-active organisms, one must consider its role in the disposition of incoming solar energy and outgoing terrestrial radiation.

SNOW AND RADIANT ENERGY

The maximum amount of radiant energy received at the earth's surface on a clear summer day, at mid latitudes, is on the order of 3000 joules/day. As the winter solstice approaches, maximum insolation decreases to less than 1700 joules/day. However, this seasonal re-

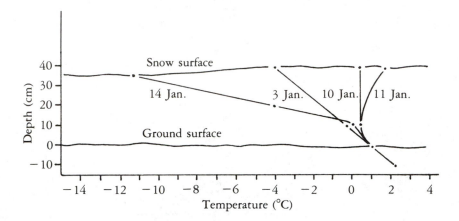

Fig. 6. *Snowpack temperatures over a two-week period in mid winter. Air temperatures ranged from −35° C to just above freezing with a rain. To the small mammal active under the snow, however, ambient temperatures remained very stable and near 0° C—a situation typical for all but extreme northern latitudes, where subnivean temperatures may be lower.*

duction in incoming solar energy tells only half the story. It is the fate of this radiation over snow-covered ground that is of greatest interest in our understanding of winter ecology.

As anyone who has spent a day on the snow in bright sunlight can attest, snow is highly reflective of incoming solar or short-wave radiation. Fresh snow is, in fact, nature's best reflector, turning back 75 to 95% of the sunlight striking its surface. This means that only a small portion of the total solar energy reaching the snow-covered ground is available to perform work in the snowpack (e.g. raise the temperature of the snow); the rest is reflected and lost back to space or absorbed by objects above the snowpack. As a snowpack ages, however, and the surface accumulates dust or bark and leaf litter, its reflectance may decrease to as low as 45%. The fate of the short-wave energy that does penetrate the snow surface will be discussed momentarily as it pertains to the light regime in the subnivean environment.

As good a reflector of short-wave radiation as snow is, it is nearly a perfect *absorber* of the long-wave or heat energy emitted by terrestrial objects. Physicists refer to an object that absorbs 100% of the energy incident upon it as a "black body," and nothing in nature so closely approximates a black body, with respect to energy in the longer wavelengths, as does a snowpack. Because every natural object emits some heat energy by virtue of its own molecular activity, the snowpack, then, is continually absorbing energy from its surroundings. The most visible evidence of this is the appearance, especially in the spring, of pronounced depressions in the snowpack around the trunks of trees, sometimes extending right to the ground surface. The tree absorbs incoming solar radiation, including some of that which the snow is reflecting; this absorption of energy increases molecular activity and raises the temperature of the absorbing surfaces; and the tree in turn emits more long-wave or heat energy. The snowpack absorbs this energy with great efficiency, which then increases the rate of sublimation around the tree trunk. The hollow is formed more by sublimation—the conversion of ice directly to vapor—than by actual melting of the snow (fig. 7).

Just as the snowpack continually absorbs heat energy, it also efficiently emits or radiates energy back to its surroundings. If our color vision extended beyond red light and into the longer wavelengths of heat energy that are invisible to us, we would see the snow surface emitting this "light" of long-wave energy. At night under a clear sky, when incoming radiation is greatly reduced, the snow surface in fact loses much more energy than it gains. Because of this energy loss the snow surface becomes colder, affecting in turn the air in contact with it, with the result that a cold layer of air often develops over the snowpack. This condition is referred to as a temperature inversion because it is a departure from the normal situation where the air close to the ground is warmer than the air aloft.

The practical point of this is that the lowest temperatures in any winter environment generally occur during the night right at the snow surface, provided there is no overhead obstruction to absorb

Fig. 7. The absorbing snowpack. Depressions in the snowpack around trees in this deciduous forest result from the absorption of long-wave or heat energy by the snowpack. The dark tree trunks absorb incoming solar radiation, the temperature of their absorbing surfaces is raised, and the trees then emit more energy at longer wavelengths. This energy is absorbed with near perfect efficiency by the snowpack, resulting in increased sublimation around the tree trunk. (Photo by Peter Marchand.)

outgoing radiation and reradiate it back to the snow. Temperatures then increase upward from the snow surface and downward into the snowpack. Figure 8 illustrates this for a northern forest clearing on a cloudless January night. It is not at all unusual to see such night-time air temperature inversions develop over the snowpack and extend several meters above the snow if the air is calm (see box on "Ice Fog"). If you were looking for an overnight shelter under circumstances like this, you would be considerably more comfortable beneath the forest canopy where long-wave energy is being reradiated downward to the snow surface. For the same reason, snowmelt is often delayed the longest in small forest clearings where the surrounding trees intercept incoming solar radiation but allow the unimpeded loss of heat energy from the snow surface to space (fig. 9).

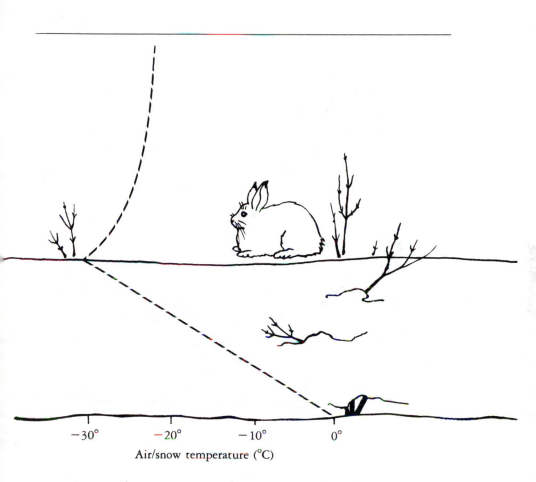

-30°   -20°   -10°   0°

Air/snow temperature (°C)

Fig. 8. Nighttime temperature profile over snow-covered ground. Air temperature is lowest on a clear night, right at the snow surface—in this case 7° colder than at a height of 2 m above the snow. The temperature increases rapidly beneath the snow surface.

ICE FOG

During the long nights in arctic regions, the snow-covered ground continually radiates heat to space, and air temperatures at the surface become increasingly colder. As long as the sky is clear and the air calm, temperature inversions continue to develop, with the height of the inversion sometimes extending several hundred meters above the surface.

a

b

Such strong temperature inversions have important implications with regard to air pollution in the North. While exhaust fumes from automobiles and emissions from smokestacks normally rise in the atmosphere because they are warmer and lighter than the air into which they are emitted, a strong temperature inversion with warmer air aloft prevents this from happening. Instead, the polluted air becomes trapped at ground level, resulting often in the formation of ice fog. At very low temperatures water vapor released from the combustion of fossil fuels freezes around other foreign matter emitted from the burned fuel to form a very dense fog consisting of minute ice particles suspended in the air. In the street scenes (*a* and *b*) photographed at midday in Barrow, Alaska, visibility in the ice fog is less than 30 m. (Photos by William Brower.)

This also explains why the coldest part of winter in the North does not coincide with the shortest days of the year—the time of lowest incoming solar radiation. It is because the snow-covered ground continues, beyond the winter solstice, to radiate more heat than it gains due to the high short-wave reflectance and efficient long-wave emittance of snow. The annual distribution of available solar energy at polar latitudes is actually asymmetrical with respect to total incoming radiation because of the presence of a snowcover that lasts well into spring. If we measure the individual components of energy exchange and subtract all the outgoing terrestrial radiation from the absorbed incoming radiation (total incoming minus reflected), we find for latitudes north of the Arctic Circle and south of the Antarctic Circle an energy deficit that persists well into the spring (fig. 10). This energy deficit is alleviated only by the occasional influx of warm air masses into these regions from temperate or tropical latitudes.

Returning to the disposition of incoming solar radiation, it is of

Fig. 9. Delayed snowmelt in forest clearings. In small forest openings, where the height of the surrounding trees is greater than the diameter of the clearing, incoming solar radiation is intercepted by the trees and very little is received directly at the snow surface. At the same time, loss of outgoing long-wave or heat energy from the snowpack is unimpeded due to the lack of overhead obstruction. The result is a negative energy balance (the snow surface loses more energy than it gains), which delays snowmelt. Increased deposition of snow in the forest opening from the crowns of nearby trees also contributes to later lying snowpacks in these situations.

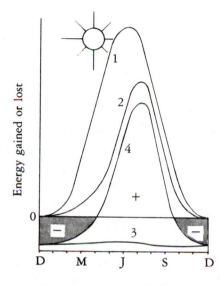

Fig. 10. Annual radiation balance at high latitudes. Curve 1 indicates the total amount of solar radiation received per month at the earth's surface, at a latitude of 70°N. Because of the persistence of a highly reflective snowcover, only a relatively small proportion of this is absorbed (Curve 2), until June when the ground is finally snow-free. For most of the winter, radiant energy emitted from the snow surface (curve 3) is greater than the total absorbed, giving a negative energy balance (curve 4), as indicated by the shaded areas.

interest now to ask what happens to the light that does penetrate the snow surface and how might its transmission through the snowpack be affected by seasonal changes in snow depth and density. Though we may be dealing with only a very small amount of radiant energy under the snowpack, its biological significance may be disproportionately high. Even very low light levels under the snow may influence a number of life processes throughout the winter. We will be considering evidence later, for example, that plants may be able to utilize light energy penetrating deep snowcover for chlorophyll production and photosynthesis. The light regime under snow may also affect small mammal population dynamics by directly influencing sexual maturation and reproductive behavior of small mammals in the subnivean environment. It is important, therefore, to have some idea of how light behaves as it passes through snow.

If snow at .2 g/cm^3 is thought of as representing one end of a density gradient, with glacier ice (.7 g/cm^3) somewhere in the middle and water (1 g/cm^3) at the other end, then one might conclude from experience that the transparency of snow increases as its density increases to that approaching ice. This has been the traditional view and there is much evidence in the literature supporting it, particularly from researchers interested in how light affects the growth of snow algae on alpine glaciers. However, if we consider our experience with snow more closely, a problem arises with this generalization. A snowball in the hand or a slab of wind-packed snow certainly appears to transmit less light (is more opaque) than a handful of loose snow held to the light, even though the density of the snowball or wind slab may be two to four times greater than the loose snow. But if our eyes are not deceiving us, how can we reconcile this observation with apparent discrepancies in the literature? The answer lies in the fact that light penetrating the snowpack behaves very differently under opposite extremes of snow density. Whether light transmission increases or decreases as the snowpack becomes more dense actually depends on whether we are dealing with a seasonal snowpack of relatively low density, typical of northern forest regions, or the highly metamorphosed permanent snowfields of arctic and alpine regions. More specifically, the behavior of light in snow depends on the degree to which the snowpack has been altered by the metamorphic processes described earlier.

The accumulation and settling of fresh snow is accompanied by a marked decrease in light transmission. This is illustrated in figure 11 for snow of two different densities. At an average density of .21 g/cm^3, just 3% of the light that passes through the surface penetrates a depth of 20 cm. Less than two-tenths of a pecent is transmitted to 40 cm. A slight increase in density to .26 g/cm^3 reduces the amount of light passing through the snow by 50% or more at any given depth.[1] This means that an increase in density of only .05

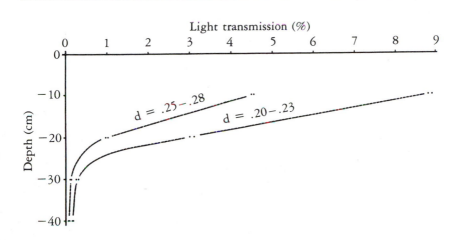

Fig. 11. Absorption of light in the snowpack. Light transmission (percent of visible light penetrating the surface) is shown here as a function of depth in snow of different densities. (From P. J. Marchand, "Light Extinction under a Changing Snowcover," in Winter Ecology of Small Mammals, *ed. J. F. Merritt, Carnegie Museum of Natural History Spec. Publ. 10 (Pittsburgh, 1984), 33–37.)*

g/cm^3 in the surface layer would have approximately the same effect as the addition of 10 cm of new snow.

The sharp reduction in light transmission after a snowfall is most pronounced at low densities. Beyond .3 g/cm^3 the effect of increasing density diminishes rapidly, and at .5 g/cm^3, maximum light absorption by the snowpack is reached. At this point the light transmission curve bends upward again, with further increases in density then resulting in an increase in light transmission as had been previously reported.[2] This turnaround in light transmission is illustrated in figure 12 for a range of snow densities from .1 to .7 g/cm^3.

It is of interest to note that the density of maximum light absorption (.5 g/cm^3) is sometimes referred to as the "critical density" of snow. This is considered to be the maximum snow density attainable by compaction alone. In order to attain higher densities, a sintering or coalescence of ice grains must occur—an effect that may

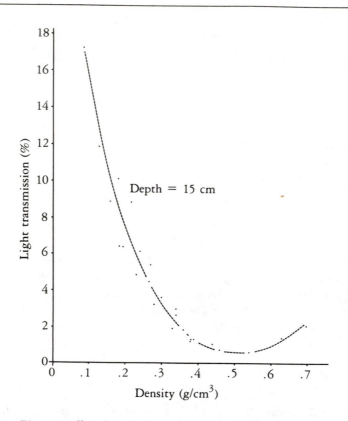

Fig. 12. *Effect of snowpack metamorphism on light penetration. Light transmission (percent of visible light penetrating the surface) is shown here in relation to snow density. (From Marchand, "Light Extinction," 35–37.)*

be achieved experimentally by warming the snow under pressure or adding water in a way similar to the processes occurring naturally over time in permanent snowfields or glaciers.

The changes in snowpack structure that take place above the critical density of .5 g/cm^3 may well explain why light transmission begins to increase at this point. Every time a light beam passes through and out of an ice grain, its direction is changed by refraction. The bending of the light beam actually occurs right at the ice-air interface. Therefore, the greater the total surface area of ice grains in a given volume of snow, the greater the amount of internal light scattering caused by refraction, and the greater the opportunity for absorption of light by the ice molecules. In the process of destructive metamorphism, increases in density occur primarily as a result of a closer packing of small grains—and more grains per unit volume means more total surface area. This increases refraction and scattering, and subsequent absorption then decreases the amount of light transmitted through the snowpack. In contrast, densities above .5 g/cm^3 are attained primarily through an increase in overall grain size due to the growth of bonds between individual ice particles. This leads to a reduction in total surface area which, in turn, reduces internal refraction and increases light transmission to greater depths (fig. 13).

How, then, is the light regime of a subnivean environment altered as the snowpack undergoes seasonal changes in depth and density? As already discussed, new fallen snow undergoes a very rapid change (destructive or equi-temperature metamorphism) as thermodynamically unstable crystals are transformed into aggregates of rounded ice grains. Within a few hours snow density may increase from less than .1 g/cm^3 to nearly .2 g/cm^3. This is followed by settling, often aided by wind, which then reduces the size of snowpack pore spaces and further increases snow density. Light transmission, therefore, will decrease sharply with time after a snowfall.

As soon as a temperature gradient is established in the snowpack, upward-moving water vapor causes a dissipation of ice crystals in the bottom layers and a growth of grains in the upper, colder layers of the snowpack (constructive metamorphism). This alone would tend to offset the decrease in light transmission, but seasonal snow accumulation and packing result in a continued increase in density through destructive metamorphism to values approaching .4 g/cm^3. A plot of light penetration with time would, thus, be asymmetrical with respect to seasonal changes in both snow-cover and incoming solar radiation (fig. 14) because the increase in incident radiation through the winter is offset by steadily increasing snow density. Less than .1% of incident light may reach the ground surface from mid January to early April in areas where snow depth is

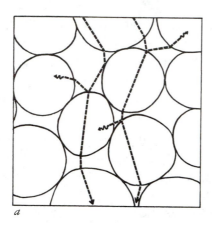

a

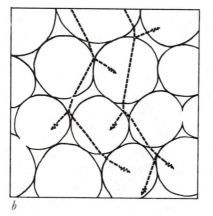

b

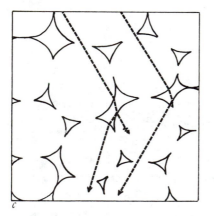

c

Fig. 13. Light scattering in the snowpack. A light beam passing through a snow crystal refracts or bends at the ice-air interface, resulting in a scattering of light within the snowpack (a). An increase in snowpack density caused by closer packing of ice grains creates more surfaces to scatter light in a given volume of space and, therefore, increases opportunity for absorption of light (b). At still higher densities, however, ice grains coalesce, reducing the amount of surface area per unit volume, and light passes through with less scattering and absorption (c). This explains the turnaround in light transmission as the snowpack undergoes metamorphism.

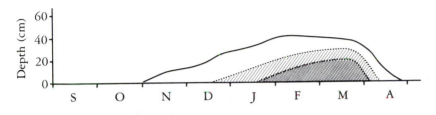

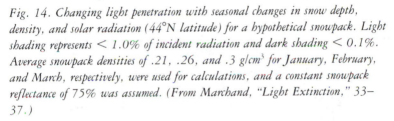

Fig. 14. Changing light penetration with seasonal changes in snow depth, density, and solar radiation (44°N latitude) for a hypothetical snowpack. Light shading represents < 1.0% of incident radiation and dark shading < 0.1%. Average snowpack densities of .21, .26, and .3 g/cm³ for January, February, and March, respectively, were used for calculations, and a constant snowpack reflectance of 75% was assumed. (From Marchand, "Light Extinction," 33–37.)

40 cm or more and average density exceeds .25 g/cm³.

Before leaving this discussion on the nature of light under snow, one other aspect of its behavior should be dealt with, and that is its quality. So far we have been dealing in a general way with "white light" or visible radiation from the sun without considering what colors or wavelengths of visible light penetrate snow most effectively. Among the best studies to date on the quality of light penetrating snow are those of Curl et al.[3] and Richardson and Salisbury.[4] Curl and his colleagues were interested in light energy as it affected the growth of algae in alpine snowfields. Using a spectroradiometer to measure light penetration of specific wavelengths under various snowpack conditions, they found that light absorption by snow was most efficient in the red region of the spectrum, with blue light reaching greater depths. A generalized graph of light penetration by wavelength from their data is given in figure 15. Richardson and Salisbury, who were also interested in plant response to light, constructed a small laboratory beneath the snowpack and used a photometer equipped with a photomultiplier tube and interference filter (which enabled selection of specific wavelengths) to measure extremely low levels of light under almost 2 m of snow. Their studies indicate that the wavelength of maximum penetration is centered

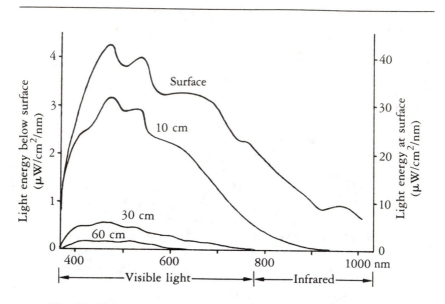

Fig. 15. Light penetration by wavelength under snow of different depths. During mid winter, snow absorbs red light most effectively, with blue and blue-green light reaching the greatest depths. (Redrawn from H. Curl, J. T. Hardy, and R. Ellermeier, "Spectral Absorption of Solar Radiation in Alpine Snowfields," Ecology 53 (1972), 1189–94.)

around that of blue-green light (about 500 μm). However, as the snowpack ages, its reflectance decreases, and as solar angles become higher, the proportion of red light penetrating the snowpack, relative to blue light, increases.[5]

Though we now know that blue or blue-green light penetrates to considerable depths in snow, we have also seen that absolute light levels under 40 or 50 cm of snow are very low at typical midwinter snow densities. Any way we measure it, it is dark down there. But that is only *our* perception. What of the organisms surviving winter under the snow? The next chapter considers whether there may, in fact, be sufficient light energy available to support some plant activity in the subnivean environment. The response of animals to such low light levels, however, is unknown and raises several interesting questions: To what extent might social interactions of subnivean mammals be influenced by changing light levels under the snow? Does light quality play a role in sexual maturation and reproductive activity in the subnivean environment, or do mammals perceive only total darkness under 40 or 50 cm of snow? Do daily (circadian) rhythms drift out of phase with "external" time-setters throughout the winter, or are the low light levels under snow sufficient to regulate daily activity patterns in subnivean mammals? These questions remain a challenge to animal ecologists. Clearly, the response of small mammals to the light conditions under snow is an area in need of much more study before any generalizations may be drawn.

3 PLANTS AND THE WINTER ENVIRONMENT

In the introductory section of this book, we considered a number of alternatives for overwintering success in the North. Our attention then was focused mainly on animals, perhaps leaving the impression that there were few evolutionary options for plants. This is not entirely the case. Some plant species have evolved mechanisms for avoiding the rigors of the winter season—for example, the overwintering seed in the case of annuals, or the below-ground corm or regenerative root stock in the case of many herbaceous perennials. Others have evolved morphological and physiological adaptations for resisting the stresses of winter. Certainly the deciduous habit of northern and midlatitude hardwoods can be considered an adaptive advantage in reducing water loss during winter when water supply is greatly limited. In the case of aspen in northern forest regions, we might even speculate that its high chlorophyll concentration in bark tissues is of adaptive value, compensating in part for the long leafless season—a possibility that will be discussed in more detail later. And the more pronounced spire form of conifers in the Far North, a result of hormonal suppression of lateral branch growth, is viewed by some as being of adaptive advantage in heavy snow country.

As was the case with animals, there are no "best answers" in an evolutionary sense for overwintering success in plants. The coniferous growth form seems overwhelmingly favored throughout the vast boreal forests of the northern hemisphere, yet there are a number of deciduous broadleaf tree species that grow right to the arctic timberline. Larch (*Larix* spp.), the deciduous conifer that combines the best of both growth forms, is capable of surviving the coldest winters on earth outside of Antarctica. Dahurian larch (*Larix dahurica*) forms the northernmost forest stands in the world, surviving in northeastern Siberia where January temperatures may go as low as $-65°$ C. In the end it is the physiology of these plants, rather than their morphological differences, that gets them through the winter. Though the snowpack offers some advantage to plants that remain under cover, protecting them from many winter stresses (and subjecting them to one or two others, as we will see later), the survival of those plants perennially exposed during winter is largely independent of the presence of snow. For these plants the major problems of winter are related primarily to low temperature stress and desiccation (the two are not entirely unrelated); the one common denominator among all trees and shrubs of northern areas is some ability to withstand these two stresses.

ACCLIMATING TO THE COLD

Acclimation is the process by which plants each year become tolerant to subfreezing temperatures without sustaining injury. It is not an avoidance mechanism whereby the plant somehow manages to escape freezing (the plant produces no "antifreeze" as such). Rather, it is a way in which the plant becomes increasingly hardy or resistant to the potentially damaging effects of tissue freezing and ice formation. In order to understand fully the process of acclimation, it is helpful first to have some understanding of how freezing occurs in the plant and what happens at the lower threshold of cold tolerance that causes permanent injury. In the following paragraphs we will consider the sequence of events that occur at the cellular level when a woody plant tissue undergoes slow cooling across the freezing point of cell sap.

Imagine an experimental setup which closely monitors the freezing of a section of woody stem, a maple twig for example, as it is cooled. In this experiment the stem section has two fine wire thermocouples attached to it, one embedded under the bark to measure tissue temperature and the other mounted about a halfcentimeter away from the bark to monitor simultaneously air temperature. With these thermocouples we can plot cooling curves for both air and stem as we subject our sample to steadily decreasing temperatures. On the basis of numerous studies of freezing in woody plant tissues, this is what we would expect to see:

At first the decrease in stem temperature closely parallels air temperature. Both descend across the freezing point of water with no divergence of the cooling curves. At a temperature of perhaps $-5°$

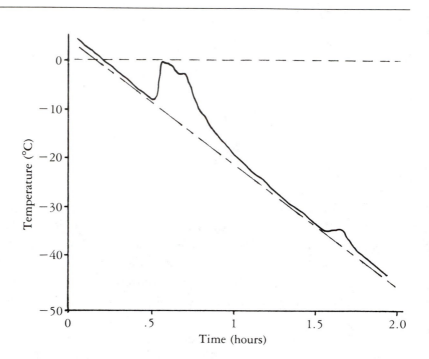

*Fig. 16. Typical freezing curve for a woody stem, in this case beech (*Fagus grandifolia*), undergoing slow cooling. The first exotherm, or rise in stem temperature, results from the release of latent heat as "flash freezing" occurs throughout the stem. The double peak indicates ice formation first in the water conducting xylem cells and large intercellular spaces and then within the cell walls and micropores of the stem. The second exotherm, occurring at a much lower temperature, coincides closely with the point of cell death and may represent intracellular freezing.*

or $-8°$ C, however, we see a dramatic departure of the two curves (fig. 16). Air temperature continues to decrease at our prescribed rate, but now the stem shows an abrupt increase to just under $0°$ C. Because of the various attractive forces with which water is held in plant tissues, the tissue is able to "supercool"—that is, cool below its actual freezing point without forming ice. As we continued to lower the temperature, though, we eventually promote ice crystallization. In this supercooled state, the initial formation of ice starts a chain reaction of "flash freezing" throughout the tissue, resulting in the release of a measurable quantity of latent heat (an "exothermic" reaction). The accompanying rise on our cooling curve is referred to as the first freezing exotherm.

All the evidence to date indicates that this first exotherm represents ice formation in pore spaces outside the living cell, a reasonable conclusion since the concentration of dissolved substances (which lowers the freezing point of water) in the intercellular spaces of the plant is lower than in the cytoplasm of the cell. The actual freezing point of this intercellular water is generally no more than one- or two-tenths of a degree below that of pure water. If our experiment is done carefully enough, two exotherms associated with this first freezing event may appear. These are separated very closely in both time and temperature, with the first representing the freezing of water in the nonliving xylem cells (the water conducting cells) and large intercellular spaces of the stem, and the second indicating freezing in micropores and cell walls.[1] The initial formation of ice outside the living plant cell is vitally important, for the process that follows this forestalls intracellular ice formation, an event that is almost invariably fatal to the cell.

Because the vibrational energy of a water molecule on the surface of an ice crystal is much reduced, an energy gradient exists between unfrozen water in the cell cytoplasm and frozen water outside the cell. As a result of this energy difference, the more active liquid molecules tend to migrate toward the ice surface. This results in an orderly movement of cytoplasmic water through the plasma membrane and out of the cell, adding its mass to the growing intercellular ice crystal. This loss of water from the cell has the effect of increasing the solute concentration of the cytoplasm and thereby progressively lowering its freezing point. As we shall see later, an important aspect of the acclimation process is an increase in cell membrane permeability so that this outward migration of water is little impeded.

As we continue our cooling experiment, the progressively smaller amounts of water freezing onto the extracellular ice crystal contribute less and less latent heat and gradually our stem sample comes close to equilibrium with air temperature again. If we push our tissue sample to the limit of its tolerance, however, we may eventually (though not always) see a second, smaller exotherm that corresponds closely with the timing of cell death.

Just what happens at this second exotherm and point of cell death is not absolutely clear, but there are at least two hypotheses that deserve consideration here. One idea put forth more than a decade ago by Soviet scientists bears merit, though it remains unconfirmed by experimental data. This hypothesis holds that at some critical low temperature an abrupt decrease in plasma membrane permeability occurs that prohibits any additional water from migrating out of the cell. The entrapped water then supercools, followed by intracellular ice nucleation and the rapid propagation of ice throughout the cell, causing the exotherm. The implication here is that the lethal damage is of a mechanical nature, with cell membranes being the primary site of injury.

Whereas there has been no confirmation yet of a sudden change in plasma membrane permeability at some low temperature, there is accumulating evidence that freezing injury to living cells is in some way associated with an alteration in resilience of the plasma membrane that may occur during extreme contraction of the membrane.[2] It is believed that the intrinsic proteins bridging the two sides of the membrane (fig. 17) may be the actual sites of damage and that once these break down they can serve as channels for the leakage of potassium ions and sugars that usually accompanies freezing injury.[3] The rearrangement of some membrane lipids into nonlayered structures upon freezing and thawing may also alter permeability of the membrane.[4]

An alternative explanation for cell death at the second freezing exotherm is the "vital water" hypothesis of Weiser.[5] This idea holds that, during freezing, a point is reached where all readily available water has moved out of the cell and is frozen intercellularly, and only water bound to structural macromolecules remains in the protoplasm. With a continued decrease in temperature, removal of

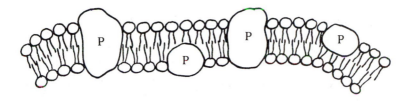

Fig. 17. *Membrane failure with cell freezing. Current models show membranes as a double layer of lipids interspersed with protein macromolecules. Some investigators believe that the intrinsic proteins bridging the two sides of the membrane are the sites of freezing injury, resulting in the leakage of cell electrolytes. Another line of thought suggests that the rearrangement of some membrane lipids into nonlayered structures upon freezing and thawing may lead to membrane failure.*

bound water initiates a chain reaction of denaturation (probably of membrane proteins among other things), additional water loss, and death by irreversible dehydration or the accumulation of toxic concentrations of solutes—the latter also capable of altering membrane properties.[6]

Whichever hypothesis proves correct, there is ample evidence that it is ice formation and not low temperature per se that directly or indirectly causes freezing injury. Even nonhardy plant tissues can survive the $-196°$ C temperature of liquid nitrogen if they are cooled slowly before immersion. This is because immersion in liquid nitrogen solidifies water without molecularly reorienting the water into an ice crystal, a process called vitrification. Just recently it has been reported that such amorphous solidification of cell contents (termed "glass formation") can occur in nature at surprisingly high subfreezing temperatures and without release of heat in a detectable exotherm. In balsam poplar (*Populus balsamifera*), intracellular fluid was observed to solidify, with slow cooling, at temperatures below $-28°$ C, effectively preventing cell injury by either intracellular ice formation or dehydration.[7] This is the first time that high temperature glass formation in a plant has ever been reported. Its possible occurrence in other species may explain why a second exotherm has

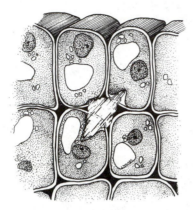

Fig. 18. Ice formation in living plant tissues. Ice crystals growing in intercellular spaces as water moves out of surrounding cells easily penetrate the fabric of the cell wall. The plasma membrane, though, remains resistant to penetration and merely deflects inward, thus preserving the integrity of the living cell.

never been seen in some winter acclimated trees and why their resistance is sometimes so extraordinary. In any event, ice outside the cell is detrimental only in the extreme case hypothesized by Weiser, where it may draw so much water out of the cell that irreversible damage occurs. Otherwise, a growing extracellular ice crystal can penetrate the fabric of the cell wall without harm, merely deflecting the plasma membrane as the membrane itself contracts with diminishing protoplasmic volume (fig. 18). This often observed phenomenon is referred to as "frost plasmolysis" and is harmless to the tissue as long as intracellular freezing does not occur.

It is important here to distinguish between slow cooling, at rates on the order of 10° C per hour that frequently occur in nature, and rapid freezing that can be accomplished in the laboratory and that occasionally occurs in nature. The difference has much to do with the survival of even the hardiest of plant species. If cooling is rapid enough that withdrawal of water from the cell cannot keep pace with the rate of freezing, then ice formation may occur within the cell and this is invariably fatal. In a study of winter damage to northern white-cedar (*Thuja occidentalis*) in Minnesota, White and Weiser found that at sunset, where the horizon was abrupt, leaf temperatures on the sunlit southwest side of the tree dropped 9.5° per minute across the freezing point of cell water.[8] This rapid freezing apparently trapped water inside the cell, where it froze and resulted in cell death. Under slower freezing rates this species was capable of withstanding temperatures of $-87°$ C without damage. It may be that much winter injury to evergreens that has been attributed to desiccation is the result of rapid freezing instead. Since freezing damage so closely resembles desiccation damage in mechanism, it is often impossible to tell the two apart by appearance.[9]

Having some understanding now of the freezing process in woody plants, it is appropriate here to ask how plants acclimate or acquire freezing resistance. This is by no means a simple question. We know from experience that the effect of low temperature varies considerably among plants; even different populations of the same species may vary in their ability to withstand freezing. We know that a light frost during the growing season may have a far more damaging effect on exposed tissue than the coldest temperatures of winter. And we know that individual plants can sometimes, by proper conditioning, be made more or less resistant to freezing damage over time. In addition to these variables we find that different tissues of the same plant may have very different thresholds of injury, and that these tolerances can change, depending on the exposure of the tissue. It is indeed a very complex situation and many questions remain unanswered today.

In spite of what remains to be learned, there appear to be some constants in the acclimation process. Freezing acclimation, the acquiring of some ability to withstand subfreezing temperatures without cell injury, begins in late summer and is ordinarily a two-step process. If you collect twig samples of a given tree species on a weekly basis, beginning in the latter part of the growing season, and subject them to a range of low temperatures in order to determine their threshold for survival, you will likely find at first that freezing injury occurs at temperatures very close to $0°$ C. Then, as the growing season comes to an end, you will begin to see an increase in their survival of subfreezing temperatures. The temperature at which freezing injury occurs will slowly drop and it will appear as if the tree will soon be able to withstand the temperatures

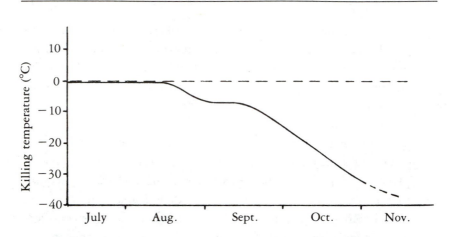

Fig. 19. *Woody plant acclimation to the cold. Freezing acclimation in woody plants normally occurs as a two-phase process. The first phase is induced by decreasing day length and results in sufficient resistance to withstand the first light frosts of autumn. Exposure to freezing temperatures then induces a second phase of acclimation that leads to deep-winter resistance typical of the species. (From C. J. Weiser, "Cold Resistance and Injury in Woody Plants,"* Science 169 *(1970): 1269–78.)*

of midwinter. However, as you continue to monitor the acclimation process, you will discover, after only two or three weeks perhaps, that the resistance of the plant reaches a plateau (fig. 19). For a while now the minimum survival temperature remains steady at around $-5°$ to $-10°$ C and it looks as if that is all the freezing resistance the tree will acquire. With time, however, the minimum survival temperature begins to decrease again as the acclimation process enters its second phase. This phase seems to be induced by the first hard frosts of the fall season, and this time the plant will reach its full midwinter freezing resistance.

The first stage of acclimation is apparently linked to growth cessation, influenced primarily by decreasing day length, but induced also by water stress. For example, the bristlecone pine attains a greater degree of freezing resistance early in the fall if active growth has been prohibited by drought during the latter part of the summer.[10] Wheat seedlings subjected to water stress for two weeks, which limits growth, and then placed in darkness have been found to develop some freezing tolerance.[11] Low but positive temperatures (e.g., 2° to 5° C) may also induce this stage of acclimation in herbaceous plants, though again the response may be related to growth cessation rather than to temperature itself.

Regardless of whether the induction factor is related to decreasing daylight, low temperature, or some other seasonal change, it is clear that the timing of phase I acclimation is subject to some preconditioning, though not all populations of a given species respond in a similar manner to the same cues. This was shown in a number of experiments by Weiser and his students using regional populations of red-osier dogwood (*Cornus stolonifera*) from several different climatic zones, which were transplanted and grown in St. Paul, Minnesota.[12] Populations from northern interior regions that normally experience very cold winters entered the first stage of acclimation as much as a month earlier than populations from milder coastal areas, though all populations eventually attained the same degree of freezing resistance.

The onset of freezing acclimation seems to involve the translocation of some biochemical compound, perhaps a simple sugar or specific hormone, which comes from the leaves, the site of perception of short days. There are a number of pieces of evidence that support this translocation hypothesis, among which are the observations that in herbaceous plants, this stage of hardiness can occur in darkness only if the plants are cold treated and supplied with sucrose; and in woody plants the leaves of hardy genetic strains can enhance the acclimation of nonhardy strains when the two are grafted together.[13] In addition, plants that are severely depleted in photosynthetic reserves do not acclimate.

The exact identity of this translocated hardiness-promoting factor remains elusive, perhaps because there may be more than one mechanism involved in the induction of freezing resistance. The plant hormone abcisic acid (ABA) almost surely plays a role in the acclimation process, at least indirectly. ABA is a photoactive hormone that is synthesized primarily in leaves and increases in concentration as day length decreases. ABA concentration also increases dramatically with even mild water stress. ABA is a nearly universal growth inhibitor, a prerequisite to acclimation, and is known to increase the permeability of membranes to water, an all-important factor in freezing resistance. [14]

In addition to the increasing ABA concentrations at the end of the growing season, a great number of other biochemical changes are associated with the acclimation process—enough, as others have pointed out, to support almost any hypothesis for freezing resistance ever proposed. Among them, changes in specific sugars and proteins, along with increasing lipid unsaturation, appear to be linked with low but positive temperature. Associated with the second stage of acclimation, induced by subfreezing temperatures, is a rapid alteration of membranes that may also involve a reorientation of macromolecules into more stable forms that can resist dehydration. For example, proteins may be converted from a polymerized configuration (many units linked together, having a high combined molecular weight) to a depolymerized state in which water is bound to each protein unit. [15] Alternatively, sugars may replace water, forming protective shells around sensitive proteins. [16]

Seasonal alteration of cellular membranes during acclimation and deacclimation is an important process both with respect to changing membrane permeability and structural integrity. In addition to the possibility that membrane proteins undergo some changes, there is ample evidence now that the lipids of at least thylakoid (chloroplast) membranes undergo pronounced seasonal alteration, too. A substantial increase in total lipid content in different age classes of Norway spruce needles closely parallels seasonal changes in ambient temperature and plant freezing resistance. Similar correlations of lipid content and freezing resistance have been reported for other herbaceous and woody species. [17] Of equal importance is the marked decrease in saturation of membrane lipids, which apparently shifts the crystallization point of the lipids to a lower temperature. [18] This has been observed in other species and may be important in maintaining higher membrane flexibility during cold periods, thus minimizing the possibility of structural damage at low temperature.[19]

It is clear that acclimation is an active process, not simply a consequence of growth cessation, and that among the many metabolic changes taking place at this time, the change in membrane permeability is of paramount importance in allowing water withdrawal and preventing intracellular ice formation. Scientists believe, in fact, that in the winter-hardy plant, the plasma membrane offers virtually no resistance to the movement of water. If this is the case, then the osmotically active substances such as sugars, organic acids, and water-soluble proteins, all of which increase in concentration during acclimation, may help to counteract the secondary effects of increased membrane permeability and protect the cell against lethal desiccation. However, the mechanism by which some species, or even different tissues of the same plant, become much more resistant than others is not known.

Even in its final state of freezing resistance, the plant may remain sensitive to fluctuations in ambient temperature. A midwinter cold wave, for example, may stimulate an increase in freezing tolerance by several degrees in as short a time as five to ten days.[20] Some researchers even recognize a third phase of acclimation that may be induced by prolonged exposure to temperatures on the order of $-30°$ to $-50°$ C and may result in a level of freezing tolerance not ordinarily attained in many species.[21] Dehardening in response to high temperatures (e.g., $15°$ C) can occur even more quickly, though rarely are midwinter thaws warm enough or prolonged enough to cause serious loss of resistance in hardy species. By the few accounts in the literature, this responsive adjustment to normal temperature fluctuations in winter may cause a shift in minimum freezing tolerance on the order of $15°$ to $20°$.[22]

Freezing injury no doubt has been a strong selective pressure in the evolution of many plant populations. Some tree species exhibit a low temperature tolerance so close to the average minimum temperature at their northern range limit (table 2) that it is safe to suggest freezing injury as the limiting factor in their distribution.

Table 2. *Tree species whose freezing tolerance closely matches the minimum temperature at their northern range limit*

Species	Killing temperature (°C)
Live oak, *Quercus virginiana*[a]	− 8
Pacific bayberry, *Myrica Californica*	− 10
Redwood, *Sequoia sempervirens*	− 15
Oregon white oak, *Quercus garryana*	− 15 to − 20
Southern magnolia, *Magnolia grandiflora*	− 15 to − 20
Slash pine, *Pinus elliottii*	− 10 to − 20
Swamp chestnut oak, *Quercus michauxi*	− 20
Sweetgum, *Liquidambar styraciflua*	− 25 to − 30
Eastern redbud, *Cercis canadensis*[b]	− 35 to − 40
Northern red oak, *Quercus rubra*	− 40 to − 41
American beech, *Fagus grandifolia*	− 41
Black cherry, *Prunus serotina*	− 42 to − 43
Sugar maple, *Acer saccharum*	− 42 to − 43
American hornbeam, *Carpinus caroliniana*	− 42
White ash, *Fraxinus americana*	− 42 to − 46
Shagbark hickory, *Carya ovata*	− 43 to − 46
Eastern hophornbeam, *Ostrya virginiana*	− 41 to − 48
Yellow birch, *Betula alleghaniensis*	− 44 to − 45

[a]A. Sakai and C. J. Weiser, "Freezing Resistance in Trees in North America with Reference to Tree Regions," *Ecology* 54 (1973): 118–26.
[b]Freezing injury in stem sections reported in M. F. George, M. J. Burke, H. M. Pellet, and A. G. Johnson, "Low Temperature Exotherms and Woody Plant Distribution," *HortScience* 9 (1974): 519–22.

Yet other species exhibit freezing tolerance way beyond any low temperature that they have or will likely ever experience (table 3), and one wonders by what mechanism they have evolved such extreme tolerance. Between these two groups there are others that show re-

Table 3. *Tree species whose freezing tolerance greatly exceeds the lowest temperature within its range (or species that show little variation in tolerance over a wide geographical area)*

Species	Killing temperature (°C)
Baldcypress, *Taxodium distichum*[a]	−30
Black cottonwood, *Populus trichocarpa*	−60
Eastern cottonwood, *Populus deltoides*	−50 to −80
Eastern hemlock, *Tsuga canadensis*	−60
Black willow, *Salix nigra*	−60 to −80
Basswood, *Tilia americana*	−80[b]
Northern white cedar, *Thuja occidentalis*	−80 ↓
Red pine, *Pinus resinosa*	−80 ↓
Jack pine, *Pinus banksiana*	−80 ↓
White spruce, *Picea glauca*	−80 ↓
Black spruce, *Picea mariana*	−80 ↓
Larch, *Larix laricina*	−80 ↓
Balsam fir, *Abies balsamea*	−80 ↓
Balsam poplar, *Populus balsamifera*	−80 ↓
Quaking aspen, *Populus tremuloides*	−80 ↓
Paper birch, *Betula papyrifera*	−80 ↓

[a]Sakai and Weiser, "Freezing Resistance of Trees," 118–26.
[b]George et al., "Low Temperature Exotherms," 519–22.

markable genetic flexibility in adjusting freezing tolerance to the minimum temperatures typical of the region in which they grow (table 4). Douglas fir growing on the coast of Oregon, where winters are relatively mild, attain a freezing tolerance to only −20° C, but the same species growing in the Colorado Rocky Mountains may attain freezing resistance to −80° C. Red maple collected during midwinter in Mississippi are reportedly resistant only to −30° C, whereas collections from Minnesota have been reported to show no injury at temperatures down to −54° C. And similar adaptation has been seen along elevational gradients. For example, Saghalien

Table 4. *Tree species that show adjustment in freezing tolerance matching geographical variation in minimum temperature*

Species	Region	Killing temperature (°C)
Western hemlock, *Tsuga heterophylla*	Coastal Oregon	−20[a]
	Idaho	−35 to −40[a]
Douglas fir, *Pseudotsuga menziesii*	Coastal Oregon	−20[a]
	Colorado Rocky Mts.	−50 to −80[a]
American sycamore, *Platanus occidentalis*	Mississippi	−20 to −25[a]
	Minnesota	−40[b]
Red maple, *Acer rubrum*	Mississippi	−25 to −30[a]
	Minnesota	−54[b]
White ash, *Fraxinus americana*	Mississippi	−30[c]
	New Brunswick	−43[c]
Green ash, *Fraxinus pennsylvanica*	Mississippi	−30 to −40[a]
	Minnesota	−54[b]
Eastern white pine, *Pinus strobus*	Tennessee	−39[d]
	Minnesota	−89[d]

[a] Sakai and Weiser, "Freezing Resistance of Trees," 118–26.
[b] George et al., "Low Temperature Exotherms," 519–22.
[c] N. L. Alexander, H. L. Flint, and P. A. Hammer, "Variations in Cold-Hardiness of *Fraxinus americana* Stem Tissue according to Geographic Origin," *Ecology* 65 (1984): 1087–92.
[d] D. M. Maronek and H. L. Flint, "Cold Hardiness of Needles of *Pinus strobus* L. as a Function of Geographic Source," *Forest Science* 20 (1974): 135–41.

fir (*Abies sachalinensis*) in the central mountains of Hokkaido, Japan, show a steady increase in freezing resistance with increasing elevation.[23] It is likely that many more species will join the list in table 4 as additional data become available.

WEATHERING THE WINTER DROUGHT

The desiccation problem

One of the most widely revered doctrines of winter ecology, particularly with reference to damage of trees at timberline, is that prevalent "dry winter winds" promote severe desiccation of plants exposed above the snowpack. At the turn of this century European ecologists were arguing that "cold winds possess extraordinarily strong drying power" which alone determined the limits of tree growth at both northern and high mountain timberline.[24] It is still thought that evergreen conifers exposed at timberline are universally subjected to damaging water loss during winter,[25] but such generalizations warrant some caution. While winter desiccation is an important stress in some locations, the conditions under which it occurs are often misunderstood. This is a case where intuition betrays us as we attempt to extrapolate human experience to the plant situation. The fact is that exposed plants face their greatest water loss problems under bright sunshine and calm conditions, the kind of winter day we perceive as most favorable. An increase in wind speed under these conditions would actually result in decreased transpiration. The logic of this often surprising conclusion is quite straightforward, but does require a careful examination of the factors involved in plant water loss.

Implicit in all arguments supporting desiccation as a major winter stress are a number of underlying assumptions. It is generally thought that even though leaf stomates, those tiny pores through which atmospheric gases are exchanged, remain closed throughout most of the winter, diffusion of water vapor through the protective cuticle of the leaf is still significant and is accelerated by low atmospheric humidity and high winds. It is also assumed that replenishment of water lost during winter is not possible, either because absorption is resticted by cold or frozen soils or because water movement is blocked by frozen stems or disrupted water columns. We will begin our consideration of winter desiccation by examining these assumptions, analyzing first the loss side of the water budget and then considering restrictions on the supply side.

The rate of water loss from a plant leaf, summer or winter, is directly proportional to the difference in water vapor concentration between the internal atmosphere of the leaf and the outside air, and is inversely proportional to the vapor transfer resistances offered by

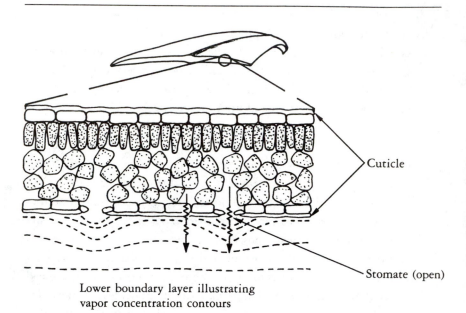

Lower boundary layer illustrating
vapor concentration contours

Fig. 20. A generalized model showing the pathways and resistances to diffusion of water vapor out of a leaf. The stomates and cuticle represent parallel pathways, with most of the water diffusing along the path of least resistance, through the open stomates. When stomates are closed during winter, water loss is restricted to diffusion through the cuticle, and leaf resistance thus becomes very high. Boundary layer resistance, which is provided by the thin shell of air surrounding the leaf, is influenced by leaf size and shape and by wind speed. It is essentially the same during summer and winter.

the leaf and the thin shell of air, the so-called boundary layer, surrounding the leaf. In other words, transpiration increases as the difference in vapor concentration between leaf and air increases, but decreases as diffusion resistances become greater. These pathways and resistances to vapor diffusion are illustrated schematically in figure 20.

The vapor concentration gradient that drives the transpiration process is strongly influenced by temperature differences between the leaf and outside air. Intercellular air spaces of the leaf are maintained at or very near 100% relative humidity as water evaporates from cell walls forming the terminus of water columns in the plant. The actual quantity of water vapor held in that internal atmosphere increases as temperature goes up. Therefore, the greater the elevation of leaf temperature above air temperature, as occurs when the leaf is absorbing direct or reflected sunlight, the greater the vapor concentration difference between the leaf and air. It is not at all unusual during periods of calm to find conifer needles heated several degrees above air temperature. In the extreme, needle temperatures 20° above ambient have been reported during the winter.[26] This is especially likely to occur during late spring when the sun is higher in the sky and there is still a snow-cover to reflect additional radiation to the exposed needle. The result might be a fivefold or greater increase in the driving force for water loss.

Counteracting this force for the outward movement of water vapor are the leaf resistances to diffusion provided by the stomates and the waxy cuticle covering the leaf. These resistances are easily measured and their units of "seconds per centimeter" (s/cm) are simply the reciprocal of the units for conductance (cm/s). The magnitude of leaf resistances to vapor diffusion are species dependent. The minimum stomatal resistance (i.e., when stomates are open) of many broadleaf species is less than 2 s/cm, whereas the stomatal resistance of conifers is considerably higher, usually near 20 s/cm. In contrast, the waxy cuticle that covers the outer layer of cells (the epidermis) is highly impervious to water. Hence, cuticular resistance to water vapor diffusion is several times greater in magnitude, generally ranging from 200 to 400 s/cm for conifers and up to 1000 s/cm for some desert succulents (see P.S. Nobel, Biophysical Plant Physiology and Ecology, for a summary of leaf resistance values).[27] Cuticular resistance has also been found to increase sharply with decreasing temperature.[28] As long as the cuticle remains intact, then, this protective layer is very effective at minimizing water loss when leaf stomates are closed, as is the case for evergreens throughout much of the winter.

An additional resistance to vapor diffusion is provided by a thin shell of air, the boundary layer, that is in contact with and influenced by the leaf. The thinner this surrounding layer, the more rapid the heat or vapor transfer through this zone because heat and vapor concentration gradients between the leaf and outside air will be steeper. This is analogous to the shell of air entrapped by the fur or feathers of an animal or by our own clothing, which also acts as a heat and vapor transfer zone. A leaf that is covered by dense hairs or has rolled leaf margins will likewise have a thicker boundary layer. The resistance this layer offers to vapor diffusion can be measured, and for the single "naked" leaf this resistance is generally less than 1 s/cm, although this will vary directly with leaf size and shape and inversely with wind speed.[29]

Wind currents are important because they reduce this boundary layer resistance. The effect of wind is twofold: it removes the more humid shell of air from around the leaf, thereby increasing the rate of transpiration; and it cools the leaf, thus tending to maintain temperature equilibrium between the leaf and air. In the latter case, the vapor concentration gradient between the leaf and air is reduced and consequently transpiration is decreased. The relative importance of these two opposing effects depends both on other microenvironmental factors and on the physiological behavior of the plant.

During the summer growth period, the most significant leaf resistance to loss of water under nonstress conditions is offered by the open stomates. Vapor diffusion through the cuticle is negligible as long as the stomates remain open. At this time of year the boundary layer resistance ($\sim$1 s/cm) is closer in magnitude to the leaf resistance ($\sim$2 s/cm) and any reduction of boundary layer thickness by air turbulence becomes significant in terms of increasing transpiration. In this case the effect of wind on leaf temperature, is less important.

In the wintertime, however, the relative importance of the diffusive resistances is changed significantly. Stomatal opening is not known to occur during the winter, as apparently it is prevented by low temperatures.[30] Even midwinter thaws do not appear to induce stomatal movement. With stomates closed, vapor diffusion occurs primarily through the cuticle, and the leaf resistance thus becomes

WIND AND THE WINTER-EXPOSED PLANT

To understand how atmospheric properties such as humidity and wind speed interact with the plant leaf to influence water loss during winter, it is useful to consider the transpiration process in simple mathematical terms. This has the advantage of allowing us to experiment with real or hypothetical numbers and to predict the effect of changing plant or environmental variables on water loss. In our model, transpiration is described by the simple equation

$$E = \frac{(c_l - c_a)}{r_l + r_a},$$

where E is the transpiration rate, $c_l - c_a$ is the water vapor concentration difference between the intercellular spaces of the leaf (c_l) and the air outside the leaf boundary layer (c_a), and r_l and r_a are respectively the leaf and boundary layer resistances to the diffusion of water vapor.

The driving force for transpiration is the gradient $c_l - c_a$, since water will tend to move from areas of high concentration to areas of low concentration. This gradient is strongly influenced by temperature differences between the leaf and outside air and, thus, is greatly affected by the heating of the leaf as it absorbs solar radiation. The influence of this heating on the vapor concentration gradient can be readily appreciated by consulting a table of saturation vapor concentrations for different temperatures. With the air at $-10°C$ and 50% relative humidity, and with a leaf at the same temperature as the air, the difference in vapor concentration between the inside and outside of the leaf would be 1.2 $\mu g/cm^3$. If the leaf were heated to $10°C$, however, the difference in vapor concentration would increase to 8.2 $\mu g/cm^3$, a substantial increase in the attractive force for water vapor moving out of the leaf. Under these circumstances, the dissipation of heat from the leaf by wind would substantially reduce the vapor concentration gradient between leaf and air.

Opposing this driving force for water loss are the diffusive resistances offered by the leaf and boundary layer of air surrounding the leaf. During the summer when leaf stomates are open and plant resistance to water loss is generally low (e.g., 2 s/cm), a reduction in boundary layer resistance from a maximum of 1 s/cm to near zero as wind speed increases will have a significant effect on the denominator of our equation. However, during the winter r_l is generally very high (e.g., 200 s/cm), while maximum r_a remains near 1 s/cm in still air. By substituting these numbers into our model, we can readily see that a reduction of r_a by wind in this case would have very little effect on the denominator of our equation. Instead, the dominant effect of wind during the winter, when diffusive resistances are very high, is to dissipate heat from the leaf, thereby maintaining a narrower vapor concentration difference between leaf and air. The net effect of wind under these circumstances would be to reduce, rather than increase, water loss.

very high, in most cases exceeding 200 s/cm. We can readily appreciate now that when leaf resistance to water loss is very high, any reduction of the boundary layer resistance to vapor diffusion by high winds is negligible by comparison, and unimportant in terms of increasing vapor transfer (see box on "Wind and the Winter-Exposed Plant"). So as long as the leaf cuticle remains intact, the dominant effect of wind in the wintertime is to force heat dissipation. This has two important consequences: maintenance of leaf temperatures below the freezing point of cell water when air temperatures are very low; and reduction of temperature differences between the leaf and air (fig. 21*a*), with consequent reduction of the vapor concentration gradient between leaf and air. In the first case, water loss is limited to sublimation from frozen tissues, requiring greater energy input than for the evaporation of free water. In both cases, the net effect of wind is to reduce, rather than increase water loss (fig. 21*b*).

The best test of these conclusions is, of course, to follow the water balance of evergreens under field conditions where their foliage remains exposed to high winds throughout the winter. This was done for the duration of two winters at timberline on Mt. Washington[31] and for a similar period near the summit of Mt. Monadnock[32] in the northern Appalachian Mountains. Neither site produced any evidence of damaging winter desiccation in healthy foliage. In fact, on Mt. Washington, the relative water content of the leafless birch stems (*Betula papyrifera*) dropped to lower levels throughout the winter than did the foliage and shoots of exposed spruce and fir (fig. 22). Furthermore, the balsam fir at timberline maintained the same foliar water content as fir growing in a Christmas tree plantation at a sheltered location near sea level.[33]

The northern Appalachian mountains are characterized by prevalent high winds and frequent cloud cover during winter, conditions which by our previous arguments would not be expected to induce significant water loss in exposed foliage. This is not so often the case in other parts of the world, where winter desiccation is reported to be a common problem. Many instances of winter drought and desiccation damage in timberline areas of Japan, central Europe, and the western United States are the result of exposed plant parts heating above freezing due to high direct and reflected radiation loads during periods of prolonged calm. Water loss under these conditions is also often enhanced by incomplete cuticle development during the shortened growing season at high elevations, and may be further compounded by abrasion of the leaf cuticle by wind-carried ice particles. High winds have other important effects on the winter-exposed plant that will be discussed later.

*Fig. 21. The relation of water loss to wind speed in dormant leaves. Maximum water loss from winter-exposed foliage occurs under calm conditions when leaves are heated strongly by the absorption of solar radiation. If leaf resistance to vapor diffusion is very high, as with an undamaged conifer needle in winter, the effect of increasing wind speed on the reduction of boundary layer resistance becomes negligible. However, an increase in wind speed does reduce leaf-to-air temperature differences (a), thus reducing the vapor concentration gradient between leaf and air. The result is a decrease in water loss from exposed foliage during winter as wind speed increases (b). (*Source: P. J. Marchand and B. F. Chabot, *"Winter Water Relations of Tree-line Plant Species on Mt. Washington, New Hampshire,"* Arctic and Alpine Research *10 (1978): 105–16.)*

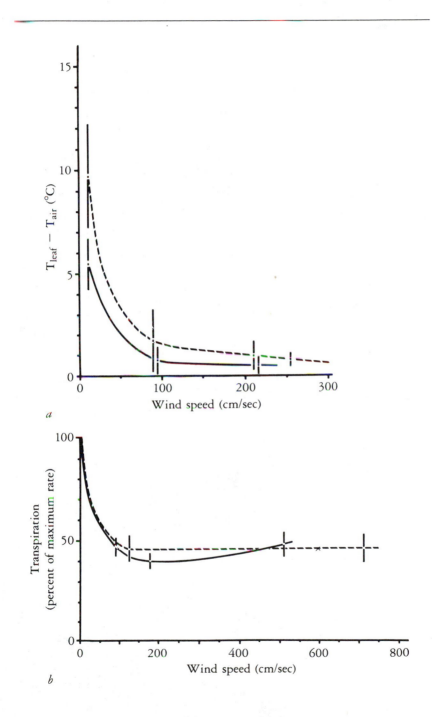

a

b

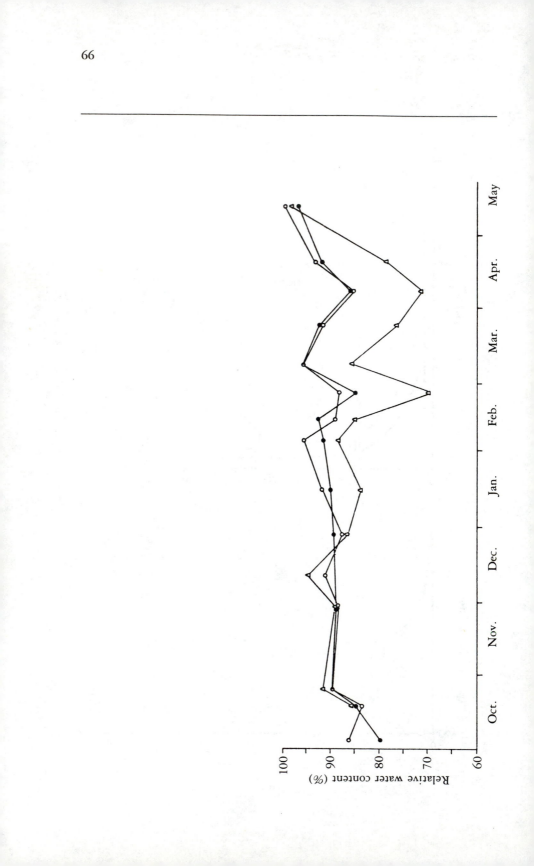

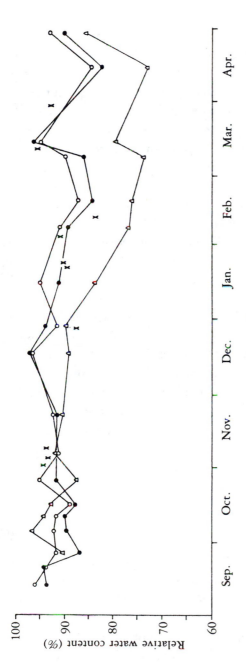

Fig. 22. Relative water content of shoots exposed to high winds throughout winter at timberline, Mt. Washington, New Hampshire. The leafless paper birch (△) exhibited the greatest seasonal decline, whereas black spruce (●) and balsam fir (○) showed the same fluctuations in water content as did balsam fir growing in a sheltered Christmas tree plantation (X) near sea level. (Source: Marchand and Chabot, "Winter Water Relations," 105–16.)

Water supply during winter

The indictment of "damaging water loss" implies that resupply of water during winter is not possible, a common assumption even among ecologists. However, a Norwegian scientist more than two decades ago calculated that the total water loss from a mature Norway spruce tree (*Picea abies*) during the course of one winter would exceed by nearly 10 times the maximum loss that it could survive if the foliage were not somehow replenished periodically. But the possibility of resupply is problematical. Apart from questions of where the water might come from, the likelihood of translocation by conventional pathways is limited by the supposed loss of continuity of the water column upon freezing. The problem is as follows:

Water movement in the xylem of a plant (the major conducting tissue) is a passive process that depends upon the cohesion of water in a continuous column within files of cells that serve as capillaries. Because of the high attraction of water molecules to one another the water column is effectively pulled through the xylem as a result of evaporation from various surfaces of the plant. This puts a tension on the water column much like that of a stretched rubber band. Normally, when the water in a capillary is under tension, as is almost always the case in the conducting elements of a tree or shrub, the dissolution of gases upon freezing causes cavitation—an air embolism—breaking the continuity of the water column, like cutting the rubber band. Without some means of forcing the column back together, there is no way to reestablish flow in the xylem of the plant.

How, then, might resupply of water to exposed shoots and foliage occur? H. T. Hammel provided one possible answer when he suggested a mechanism by which freezing in the xylem of conifers could happen without cavitation.[34] Water moving in the xylem of a conifer must pass through bordered pits located at the ends of each conducting cell (fig. 23). Stretching across each bordered pit is a fibrous membrane with a thickened central portion known as the "torus." Water moves freely around the torus as long as the torus remains centered in the pit. However, if anything causes unequal tension or

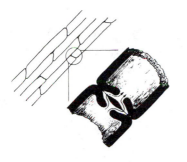

Fig. 23. Diagrammatic illustration of a pair of bordered pits between two conducting cells (tracheids) in a conifer xylem. The pits are separated by a fibrous pit membrane with a thickened central portion called the "torus." The torus acts as a valve, closing tightly against one side or the other of the bordered pit in response to small pressure changes in the xylem.

pressure distribution on either side of the torus, it deflects toward the side of highest tension, closing off the bordered pit. A pressure drop of as little as .03 MPa (5 psi) across the torus may be all that is needed to activate this check-valve.[35]

According to Hammel, this is just what happens when freezing occurs in the xylem of conifers. Because of the expansion of water upon conversion to ice, the freezing of only 1% of the water confined in a tracheid (the basic conducting element in the xylem) would reduce hydrostatic tension in that cell by about .2 MPa or 30 psi. Since there is likely to be only a slight tension in the water column at the time of freezing (atmospheric "pull" is normally low then), only a small fraction of the total water need freeze before all the tension is relaxed and the torus closes tightly. As the remaining water freezes and expands, hydrostatic pressure then builds inside the relatively rigid xylem cell, possibly reaching as high as 6 MPa, or more than 900 psi. The resistance of the cell wall to expansion means that the air bubbles now frozen out of solution will be under tremendous pressure as soon as thawing occurs, forcing them to redissolve almost immediately upon warming.

With thawing, pressure changes in the cell follow the reverse pattern. As the ice melts the pressure is gradually reduced to zero and then, with the final fraction of ice melting, the original tension gradient is reestablished and the torus relaxes to its original position. The continuity of the water column has been restored.

The key to this mechanism for preventing cavitation is confinement of water within a closed conducting element so that sufficient pressure will build with ice expansion to force resolution of dissolved gases after thawing. However, in the broadleaved trees and shrubs, the architecture of the xylem is quite different and no means of confinement is available. The cells that make up the bulk of the conducting tissue in these plants simply have perforated end walls, like sieves, through which water moves freely, with no anatomical feature at all analogous to the torus of the conducting cells in conifers. In fact, in "ring-porous" species (those that produce very large diameter conducting cells in the early part of the growing season, as opposed to the uniform small diameter conducting cells of "diffuse-porous" species), the end walls of the conducting cells are completely dissolved away so that water moves in long open-ended conduits. When water freezes and expands in these cells, it simply forces the unfrozen water to flow away from the freezing zone, relaxing tension in the water column slightly, but without ever creating the positive pressure needed to force resolution of air bubbles, even after the whole column is frozen.

The emptying of vessels upon cavitation of the water column may help explain why ring-porous trees, like the oaks and hickories, are absent from boreal forest regions where winters are so long. But what about all those diffuse-porous species that are so successful in the North? They differ only in having much smaller diameter vessels; they also lack any toruslike mechanism for isolating water in separate xylem elements. Looking again at the data in figure 22, we see periodic increases in the water content of the diffuse-porous paper birch (*Betula papyrifera*) during winter, which parallels that of the two conifers. How might this be accomplished?

For some illumination on this problem, we can look at the early experiments of J. R. Havis.[36] Using a specially constructed cooling collar placed around shrub stem segments, Havis studied the movement of water in mountain laurel (*Kalmia latifolia*) and American holly (*Ilex opaca*) during freezing. By cooling a stem through the first freezing exotherm and then applying acid fuchsin dye solution to the cut end of the stem, he was able to demonstrate water movement through the wood even though the xylem was blocked by ice. He was also able to show quantitatively that the rate of water movement did not slow until the stem was cooled through a second freezing plateau, at which time all flow then stopped. Havis interpreted this to mean that the primary pathway of water movement in the stems of these dormant plants was through the interconnected cell walls rather than through the conducting vessels themselves. If that is the case, this may circumvent the problem of cavitation. In this way, plants maintain some supply of water during winter whenever temperature conditions allow, until the water column can be restored through renewed cambial divisions in the spring.

Havis was not the first to demonstrate dye movement in stems under subfreezing temperatures. G. Hygen in Norway used a similar tracer under field conditions to make a case for resupply of water to drying foliage of Norway spruce.[37] Injecting dye into a tree just above the snow surface, he found that water moved upward during mid winter at a rate of 3% of the summer flow. Thirty years earlier, a fellow Norwegian, N. Polunin had also reported that dye moved through roots of silver birch (*Betula adorata*) imbedded in solidly frozen soil.[38]

With the increased use of radioactive tracers and heat pulse techniques in the late 1960s, several more attempts were made to measure water uptake and conduction in the field during winter, usually with the same result. Water movement was noted whenever stem temperatures were above freezing, even though soils were still frozen. In one of the more detailed studies of its kind, P. Owston and his colleagues followed water uptake seasonally in lodgepole pine (*Pinus contorta*) and red fir (*Abies magnifica*) by introducing radioactive phosphorus (^{32}P) to the clipped ends of lateral roots and then monitoring the ascent of the tracer with a portable scintillation counter.[39] During winter they found movement in only two of five red fir and none in the five lodgepole pines injected. In this case they attributed the lack of water conduction to the freezing of exposed lateral roots. In April, which was the coldest month of their study period, they found only one radioactive path, again in red fir, for 10 new injections made at the beginning of the month. However, the tracer that was injected in April began moving in all trees in early May, even though the snow was still 3 m deep and the soil temperature remained around 1° C. This suggests that it was low temperatures in the bole above the snowpack rather than low soil temperatures that was restricting water movement.

Whether or not water absorption from frozen soils is possible remains problematical. The presence of dissolved solutes may lower the freezing point of soil water a fraction of a degree;[40] and for the same reason that water in a plant may supercool, some fraction of the soil water may also remain in liquid form below the freezing point. It is possible that retention of water on soil particles with fairly high tensions may preserve a liquid film around the particle, while free water in the soil pore spaces is frozen. As long as the tension in the water column of the plant root is greater than that in the soil, then this unfrozen water should be available to the plant. Even when soils are not frozen, however, the mobility of soil water during winter is restricted by low temperatures. The viscosity of water at 0° C is twice as great as it is at 20° C. Nevertheless, the slow migration of water toward plant roots may be sufficient to keep pace with foliar losses at this time of year.

The fact that conduction may resume whenever aboveground tissues thaw suggests another source of water that may be available to drying foliage. That is the water stored in the plant itself; water held in conducting cells, the cell walls, and intercellular spaces of the tree or shrub. It has been estimated that the sapwood of a tree, generally representing 20 to 40% of the radial dimension of the tree, holds enough water to satisfy all its requirements at a moderate transpiration rate for one week.[41] At lower transpiration rates the supply would, of course, last longer. With cuticular water loss from dormant conifer foliage being some 40 times less than that of actively transpiring foliage, water stored in the sapwood might well be sufficient to carry the tree through the whole winter.[42] This assumes that opportunity for mobilization of water occurs with sufficient frequency throughout the winter. Given that conifer needles are often observed to be above freezing even when ambient temperatures are lower, it seems reasonable to expect that exposed parts of the bole and branches would likewise be warmed. It is likely, then, that the upward movement of water reported in some conifers during winter represents internal redistribution rather than actual water loss from foliage at the time of measurement.

One other possibility remains for replenishment of water, and that is through foliar absorption of water vapor. There is little question that plants that remain buried in the snowpack all winter gradually make up any water deficit with which they may have entered the season. As long as the snowpack air spaces are saturated with water vapor and the plant beneath the snow remains at the same temperature as the snowpack (i.e., does not heat by absorption of light penetrating the snow), then the direction of water vapor transfer will always be from the air to the leaf. However, for the tree or shrub exposed above the snowpack, opportunity for foliar absorption may be very limited, even where branches are laden with snow. Even a leaf under some water stress will have an internal atmosphere of greater than 99% relative humidity, and it is not often that ambient humidity will exceed this. Nevertheless, there remains the possibility that as long as some portion of the plant's foliage lays buried under the snowpack, enough absorption may occur to benefit exposed foliage through internal redistribution of water.

Recently, my students and I attempted an experiment to determine the relative contribution of these various pathways to resupply. While not conclusive yet, the results to date have been interesting and provide some insight into this problem. In our study, three treatments were designed to eliminate systematically one pathway of water absorption or another in red spruce (*Picea rubens*) and balsam fir (*Abies balsamea*) trees, each about 3 m in height. The treatments, described below, were initiated in January after a snowcover of approximately 1 m had accumulated, burying a substantial portion of the foliage of these trees.

In the first treatment we carefully cut trees at their base with a long-handled pruning saw while the snowpack was left as undisturbed as possible. The base of the tree was sealed by inserting a jar cap filled with melted parafin between the cut surfaces. The trees were then guyed with rope while the lower branches remained buried in the snow. The intent of this treatment was to elimate root absorption while still allowing for the possibility of foliar absorption and internal redistribution of water.

In the second treatment, the trees were left intact to allow for root uptake, but all branches below the snow were pruned to eliminate any foliar absorption. This resulted in the removal of about one-third of the foliage of each tree, but did not alter the total amount of foliage exposed to water loss above the snowpack.

In the third treatment, we made no alterations to the tree itself, but we removed all snow from around the tree and kept the ground clear for the remainder of the winter. This resulted in freezing the soil to considerable depth. The intent here was to minimize root uptake and translocation while at the same time eliminating any possibility of foliar absorption.

For the duration of the winter we monitored the water content of exposed foliage on all trees, including undisturbed control trees, weekly. Each winter that we conducted this experiment the results were the same, typified by the data shown in figure 24. Throughout the winter no treatment produced any substantial difference in foliar water content. Exposed foliage of the experimental trees periodically gained or lost water in the same manner as the control trees—until very late spring. As air temperature warmed and snowmelt accelerated, all treatments except one showed a pronounced increase in water content. At this point the basally cut trees experienced a dramatic decrease in foliar water content.

Until late spring the common denominator which kept the foliage of all trees fully hydrated was apparently stored reserves. Regardless of treatment, all trees appeared to have sufficient water available internally to carry them through the winter. Only when internal water was exhausted and the basally cut trees had no means of uptake did the difference due to that treatment appear. While

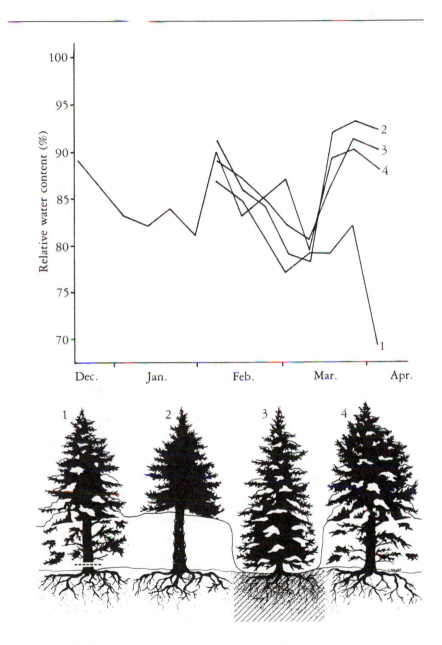

Fig. 24. *Water content in exposed foliage of red spruce,* Picea rubens, *following the treatments indicated.*

treatments designed to show foliar absorption or root uptake from frozen soil throughout the winter did not result in increased foliar water content, it is clear that resupply of water to these other trees was possible sometime before the basally cut trees exhausted their internal reserves.

It should be reiterated at this point that a plant frequently exposed to high direct and reflected solar radiation during periods of prolonged calm may have a problem maintaining a favorable water balance during a long winter. Under these circumstances any opportunity to absorb water, whether by roots or foliage, may be advantageous. It is clear that in some situations, most notably at timberline in the European Alps and in some mountain ranges of the western United States where stunted trees may have limited internal reserves, severe desiccation may be an important factor in the overwintering success of plants.

THE EVERGREEN ADVANTAGE

We have seen that retention of foliage through the winter may prove, under certain conditions, to be something of a liability. If this is the case, then why be evergreen?

It is often supposed that the advantage of the evergreen habit in conifers is to enable continuance of photosynthesis throughout the winter. Since chlorophyll in these trees does not undergo the seasonal degradation so obvious in the autumn leaves of deciduous trees, it is presumed to remain functional in its light-trapping capacity—and green is equated with photosynthesis. However, it is probably safe to say that midwinter photosynthesis in northern conifers is more the exception than the rule, although there is little doubt that the photosynthetic season may be extended considerably by the evergreen habit.

Whether or not photosynthesis occurs at all in exposed conifer needles during the winter seems to depend as much on external factors as on any internal plant cycles. In milder winter climates a number of conifers have shown small positive net photosynthesis whenever temperatures climb above freezing. Austrian pine and Norway spruce at low elevations in central Europe have been found to photosynthesize on warm winter days at rates sometimes equalling 25% or more of their summer and fall rates. In the subalpine forests of New Zealand's Southern Alps, some evergreen trees show only a brief photosynthetic dormancy. There the occurrence of daily frosts in late autumn (June) is accompanied by a complete cessation of photosynthesis in mugo pine, mountain beech, and lodgepole pine. Barely two months later, though, while nighttime temperatures are still several degrees below freezing, this dormancy is broken. Photosynthesis measured when daytime needle temperatures approach 12° to 15° C can be as much as 10% of maximum late summer rates for mountain beech and 35% for the pines. Still higher photosynthetic rates have been recorded during late winter at lower elevations.[43]

In more severe winter climates, the situation is very different. At elevations near timberline in the European Alps, the photosynthetic machinery of Austrian pine appears completely shut down during midwinter and the trees are not at all able to take advantage of brief winter thaws.[44] Likewise, with bristlecone pine growing near timberline in the White Mountains of California, no photosynthesis occurs during winter even though daytime temperatures may rise above freezing for several days in any winter month. The carbon balance (photosynthetic uptake of CO_2 minus respiratory loss) of these trees remains continuously negative from mid November to the end of April.[45]

Indirect evidence indicates that the photosynthetic machinery of other northern conifers may also be inactivated during the winter months. In balsam fir, the chloroplasts of leaf mesophyll cells (beneath the epidermis) are clumped together during midwinter rather than being distributed evenly around the perimeter of the cell for maximum light-trapping efficiency, as they are in the summer months.[46] In black spruce, the leaf stomates through which the CO_2

for photosynthesis diffuses show sensitivity to changing light levels (itself a photosynthetic response) only after preconditioning for several days at constant above-freezing temperatures.[47]

Even where conifers are not able to take advantage of brief winter thaws, their photosynthetic machinery may be reactivated surprisingly early in the spring. Mesophyll cells of balsam fir begin to show signs of structural reorganization, including chloroplast realignment, two months before bud break. Even at timberline pronounced increases in needle starch accumulation as a result of photosynthetic activity can appear in March, long before winter shows any signs of yielding to spring. This suggests that the photosynthetic process is not coupled to any of the metabolic processes controlling acclimation and freezing resistance. It seems reasonable to conclude that evergreen conifers benefit, at the very least, from an extended photosynthetic season, if not a continous one.

What, then, might we expect with other evergreen plants, particularly those that remain under the snowpack throughout the winter? With regard to temperature, subnivean plants enjoy a much more benevolent environment—one which is also free from problems associated with water stress. We know that many plants remain or turn green and actively grow under the snow, sometimes even forming flowers before snowmelt. In many cases this is accomplished through the utilization of stored energy reserves; but is it not possible that some growth is supported by photosynthesis under the snow? Although light levels under the snowpack are much reduced, the light that penetrates to greatest depths is of a wavelength near optimum for chlorophyll absorption. Chlorophyll synthesis has been measured in plants responding to the light penetrating as much as 80 cm of snow. Furthermore, the elevated CO_2 levels sometimes found under the snowpack (a point that wil be discussed later) could reduce the light-compensation point such that positive net photosynthesis might be possible under lower light levels than normally expected.

Too few attempts at measuring photosynthesis in subnivean plants have been made to allow any generalizations yet, but the evidence points to at least limited photosynthetic activity under the snow. L. Tieszen was able to measure limited carbon fixation in three arctic herbaceous species just before snowmelt, but also found that the plant leaves lacked their full complement of enzymatic activity, as is often the case with immature leaves, and this reduced their photosynthetic potential.[48] In the laboratory, Erici Mäenpää simulated the light and temperature conditions that prevail under 30 cm or more of snow during the springtime in northern Finland and succeeded in measuring positive net photosynthesis in the green stems of bilberry (*Vaccinium myrtillus*) under those conditions.[49] And in Austria, W. Tranquillini reported "break-even" CO_2 assimilation in Austrian pine seedlings under 50 cm of snow.[50] It is clear from these examples that some, perhaps even the majority, of the subnivean plants that remain green through the winter have the ability to utilize light energy penetrating the snow.

Plants that retain their green leaves during the winter may not be the only ones to benefit from an extended photosynthetic season. More than 50 tree species, the majority of them deciduous, are known to have chlorophyll in their bark tissues.[51] In aspen, the most studied of these species, the bark may comprise up to 15% of the total photosynthetic surface area and may contribute substantially to the annual carbohydrate gain of the tree. Photosynthesis in aspen bark has been detected in winter, and it is speculated that this may enable this species to compete with the northern conifers during the long leafless season. In fact, winter photosynthesis in bark tissues may be the rule rather than the exception. T. Perry has measured photosynthesis in leafless twigs of sycamore, oak, and pecan immediately after thawing.[52] Similarly, Mäenpää in Finland has found evidence of photosynthesis in the green stems of bilberry under subnivean conditions, as noted earlier. A substantial amount of chlorophyll has also been found in the bark tissues of larch, and it is interesting to speculate as to whether or not winter photosynthesis might compensate for the deciduous habit of this conifer as well.

It seems reasonable to assume that winter photosynthesis in bark tissues operates under the same constraints as in evergreen leaves and thus would vary with the length and severity of the winter season. There is, however, one important difference: the resistances to diffusion of CO_2 that limit photosynthesis in leaves would not be the same in bark tissues. In fact, much of the CO_2 supply to the bark chloroplasts may come from internal sources. It has been estimated, for example, that in aspen 50 to 75% of the CO_2 released by respiration is re-utilized in bark photosynthesis.[53] If this is typical of other species as well, then it could make a significant difference in the amount of winter photosynthesis taking place.

Whether or not the products of photosynthesis in bark tissues are transported to any other parts of the tree remains a question because of the separation of these outer layers from the phloem tissue, which is the major transport system for nutrients in the woody plant. It may be that carbohydrates produced by the bark choloroplasts are utilized primarily in the outer layers of the plant. Nevertheless, any opportunity for photosynthesis, however limited, would help offset the continuous respiratory loss of carbohydrates in woody plants during the winter and would thus be of some adaptive value in northern areas.

MECHANICAL PROBLEMS: THE BRUTE FORCES OF WINTER

The mechanical aspects of the winter environment—heavy snow loads, blowing ice particles, even the browsing activity of mammals—pose a number of problems for plants. The effects are often direct and readily observable, but at times may be indirect and subtle, becoming apparent only sometime after the damage has occurred. In some places these forces have a profound influence on the distribution of plant communities and the development of plant growth form.

Snow loading is always a potential hazard to trees and shrubs, especially in boreal forest regions where winters are characterized by periods of prolonged calm and snow remains on branches for an

Fig. 25. Burdened by winter. Snow loading poses a particularly severe problem to conifers in regions where snows are heavy and where periods of prolonged calm result in long retention by the tree. In northern Finland, where winter storms are fed by moisture from the Baltic Sea, tree breakage under extreme snow loading, as shown here, is considered to be the major limiting factor at treeline. (Photo by Kari Laine.)

extended time. In northern oceanic climates where milder winter temperatures and high moisture result in more frequent ice storms and heavy falls of wet snow, repeated bending and breaking of trees is common (fig. 25). The pronounced spire form of conifers in northern regions helps minimize snow loading, but the weight of snow retained in some cases is still impressive. In the spruce uplands of Finland, where winter precipitation is influenced by the Baltic Sea, a tree 12 m in height can accumulate a mass of snow and ice nearly 50 cm thick and weighing 3000 kg! So dominant a force is this in the hill country of southern Lapland that tree breakage there is considered to be the major factor limiting forest stand development at timberline.[54]

In areas continually exposed to high winds, it is not heavy snow loads, but rather the abrasive force of blowing ice particles that shapes the forest stand. Just above the snow surface, the stinging force of ice particles whipped by the wind is particularly strong, and its effect is obvious wherever objects protrude above the snowpack. A wooden post covered with several layers of paint and protected on top with an inverted can provides graphic evidence of the abrasive force of this ice blast (fig. 26*a*). Likewise, bark on the windward side of exposed trees is deeply pitted, even completely worn away (fig. 26*b*). Foliage is stripped, sometimes cleanly and sometimes only partially, leaving many broken needle stubs (fig. 26*c*). In some cases ice blast removes foliage already damaged by desiccation or freezing injury; but in more severe situations healthy foliage and bark are abraded away by the force of wind-carried ice particles.

With damage to fresh tissue, water loss may then be accelerated, further aggravating the mechanical injury.[55] The initiation of wound-healing processes, including callous tissue formation, is apparently delayed during winter, resulting in reduced water content of affected branches due to the exposure of open-ended vascular tissue (fig. 27). With continued drying, the loss of foliage may be compounded. If this desiccation is not lethal, the branch may rehydrate the following spring. The reduction of photosynthetic capacity accompanying such injury, though, may be an important factor hindering the recovery of conifers because much of the shoot growth during the early part of summer is dependent upon the starch reserves accumulated in the older foliage during late winter and spring.

Fig. 26. *The abrasive force of blowing ice. On a windswept slope in the Southern Alps of New Zealand, blowing ice particles abrade through several layers of paint into the wood of this post where it is not protected by the inverted can* (a). *The same abrasive force dramatically sculpts the boles of trees exposed to years of ice blast in the Rocky Mountains* (b) *and strips healthy foliage from exposed conifers at timberline in the White Mountains of New Hampshire* (c). *(Photo credits: 26a and 26c by Peter Marchand; 26b by Friedrich-Karl Holtmeier.)*

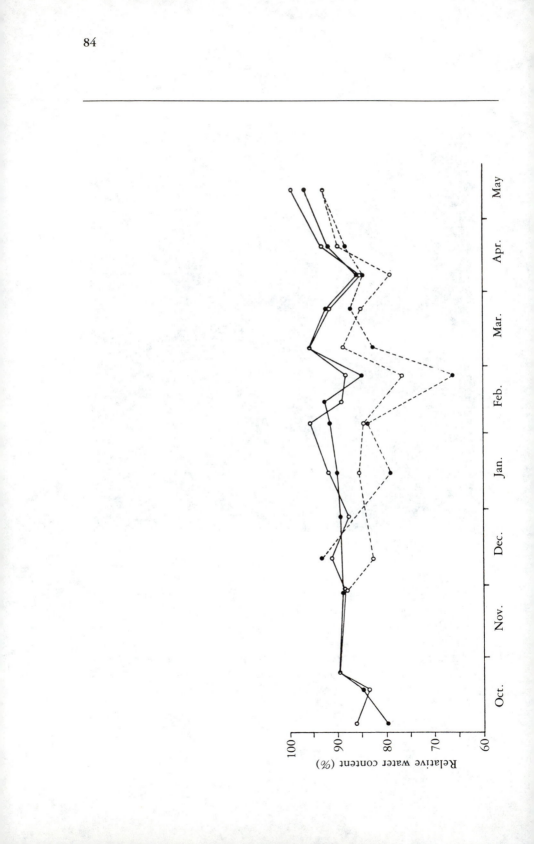

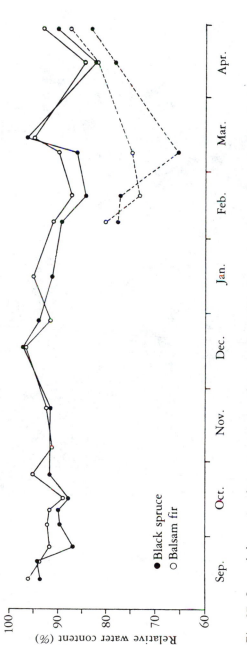

Fig. 27. Seasonal changes in relative water content of branches from which needles have been stripped or broken (dashed line) compared with the water content of undamaged foliage (solid line). Exposure of open-ended vascular tissue leads to increased drying and this may further compound the loss of foliage. Tissue that is not permanently damaged by desiccation, however, usually rehydrates quickly in the spring. (Source: Modified from Marchand and Chabot, "Winter Water Relations," 105–16.)

a

b

Fig. 28. Tree forms shaped by snow and wind. Mutual protection and increased snow packing in the downwind direction result in the growth of wedge-shaped islands or "hedges" (a) on exposed slopes in the Rocky Mountains. "Table trees," this one in Alaska (b), form where lower branches remain protected under the snowpack, but wind-carried ice particles completely remove foliage just above the surface of the snow. When trees grow through this zone of maximum abrasion,

c

d

*they sometimes take on a "mop head" form, a process referred to as
"broomsticking," exhibited by trees on this wind-exposed ridge of Mt.
Washington, New Hampshire (c). Some species, such as black spruce (Picea
mariana), have sufficient genetic flexibility that they may grow completely
prostrate under extreme environmental pressures (d), thus finding complete
protection even under very shallow snowcover. (Photos by Peter Marchand.)*

The outward effect of this abrasive damage is the development of several distinct growth forms, usually prominent near timberline where trees are most exposed. Wedge-shaped islands or "hedges" form where the windward edge of a group of trees is severely damaged by either extreme desiccation from exposure to strong sunlight or by ice blast, but where mutual protection of branches and greater snow packing to the leeward side result in increased growth in the downwind direction (fig. 28*a*). "Table trees" occur when foliage survives beneath the snow but is almost completely abraded away at the snow line. Where some protection is afforded to the terminal bud by a surrounding cluster of lateral buds, vertical growth may persist until the tree eventually reaches above the zone of abrasion and retains a healthy top (fig. 28*b*). "Broomsticking" is the term sometimes used to describe this process that results in a branchless, windswept trunk with only a "mop head" cluster of foliage at the top and a less pronounced "table" of foliage at the base. (fig. 28*c*). Mat forms develop where species with sufficient genetic flexibility, such as black spruce, grow completely prostrate under conditions of extreme exposure, without ever producing a vertical leader, and are thus completely protected by very little snow (fig. 28*d*).

In a problem closely related to ice abrasion, the frequent deposition of rime ice at high elevations, accompanied by strong winds, also results in heavy wintertime foliage loss. Rime ice forms when supercooled cloud droplets freeze upon impact with cold surfaces as clouds sweep across mountain slopes. Resulting accumulations of ice, building on the windward sides of all exposed objects, sometimes form broad featherlike vanes of great intricacy and beauty (fig. 29*a*) that wreak havoc as winds shift and break the ice-encrusted foliage. The preponderance of green litterfall in coniferous forests under this influence usually becomes evident in the springtime as litter appears at the surface of the melting snowpack. Litter traps placed under the snowpack in subalpine balsam fir stands indicate that as much as 20% of the foliage of trees exposed at the edges of canopy gaps may come down during the winter (fig. 29*b*). For trees growing in a marginal environment and under stress from other

a   b

Fig. 29. Rime ice accumulation and litterfall in subalpine forests. Supercooled water droplets freeze on impact as clouds sweep across cold mountain slopes, encasing exposed trees in glistening ice (a). Shifting winds break the encrusted branch tips, resulting in substantial loss of green foliage during the winter (b), foliage which would later have supported new growth of the tree. (Photos by Peter Marchand.)

factors, this loss may pose a significant problem and appears to be a major factor in the death of trees in subalpine fir forests of the northern Appalachians.[56]

A less destructive result of snow drifting and deep accumulation in many subalpine areas of the Rocky Mountains in the United States is the development of rather striking "ribbon forests." These forests are characterized by long bands of trees running perpendicular to the direction of the prevailing winds and separated from each other by wide corridors of moist subalpine meadow. Snow is a causal factor, but its influence in this case is on tree reproduction rather than destruction of existing stands. Trees growing in the protection of a ridge line may accumulate huge drifts of snow in its lee, snow that has blown off the exposed alpine tundra upwind. These drifts, sometimes 6 m or more in depth (fig. 30), may persist through

Fig. 30. Forest "ribbons" lying perpendicular to the direction of prevailing winds at 3350 m elevation on the east slope of Mt. Audubon, Rocky Mountains, Colorado. Zones 1, 4, and 7 depict stands of wind-flagged Engelmann spruce and subalpine fir that accumulate huge drifts of snow in their lee (photo inset). These are separated by "glades" (Zones 2–3 and 5–6) in which tree seedlings are uncommon because of the very short snow-free season. (Drawing courtesy of F. K. Holtmeier. Photo by Peter Marchand.)

almost the entire following summer. Wherever the melting edge is located in late June or early July, soil temperature and moisture conditions may be optimal for the establishment of tree seedlings, which grow to form the next ribbon downwind. These seedlings will be protected under the drift again during the following and subsequent winters, until they are tall enough to stand above the snow and begin to influence the process themselves. Where the melting snow exposes open ground much later in the growing season, only the herbaceous meadow vegetation is able to establish itself, and so a snow glade forms and is maintained by the perennial drifts as long as the forest ribbon upwind remains intact. Disturbance caused by a fire or "blowout" in the natural snowfence will change the snowdrift pattern and may then result in a shift in the ribbon and glade pattern, with tree seedlings becoming established in the glade.

4   ANIMALS AND THE WINTER ENVIRONMENT

L. WALKER

Birds and mammals that remain active throughout winter in the North face a special problem: to survive they must maintain a body temperature within quite narrow and relatively high limits. This means that the animal must produce enough heat by metabolism of its food or fat reserves to offset that which is lost to its cold surroundings. At a time when mobility is restricted and food resources are scarce, survival of the cold often becomes equated with heat conservation. To appreciate fully the problems that homeotherms face at extreme low temperatures and the energetic advantages of their many physiological and behavioral adjustments during the wintertime, we must understand the details of how heat is exchanged between the animal and its surroundings. This will require consideration of how the various modes of energy transfer—conduction, convection, radiation, and latent heat exchange—are important to the animal wintering in the North.

THE BASICS OF ENERGY EXCHANGE

The first of these energy transfer processes, heat exchange by conduction, involves the transfer of energy through molecular collisions. This requires intimate contact of the media involved in the heat transfer—for example, between skin and air or the foot surface and ground (in the case of an animal). Heat transfer by conduction is, therefore, very much dependent upon the total surface area of exposure. The rate of heat loss by conduction also depends upon the efficiency with which heat moves through the conducting medium and is influenced by the temperature difference between the heat source and heat sink as well. In general, high density materials have a high thermal conductivity, transferring heat much more rapidly than low density materials. The earlier discussion of heat transfer in a snowpack revealed that the density and thermal conductivity of a porous material like soil or snow often varies according to water content or the amount of air-filled space in the material. The same is true of an insulating layer of fur or feathers. Thus, the wetness of an animal's fur, for example, or the amount of air entrapped in the fluffed feathers of a bird will greatly affect its comfort at low temperature. Apart from material differences in efficiency of heat transfer, the temperature gradient through the transfer zone will also influence the rate of heat exchange. When two objects have greatly different temperatures, the rate of heat flow by conduction from one to the other will be much faster compared to when the two objects have nearly the same temperature.

Given these variables it is possible to estimate the amount of heat loss by conduction from an animal in a given situation. We must know only something about its total surface area of exposure, the thermal conductivity of its skin and fur or feathers (i.e., its thermal transfer layer), and the temperature of its surroundings. These variables are related mathematically by the simple equation

$$Q_c = kA \frac{(T_b - T_a)}{d},$$

where thermal conductivity (k) is measured in units of W/cm °K, surface area of exposure (A) is measured in cm^2, and the temperature gradient between the animal's core (T_b) and outside air (T_a), divided by the distance (d) in cm between the two measurements, is recorded in °C (or °K), giving us total heat loss (Q_c) in watts. We will see momentarily how helpful this model is in understanding an animal's behavior at low temperature.

The second mode of energy transfer, heat exchange by convection, involves the transfer of energy via a moving fluid, specifically air or water. In this mode, heat is gained or lost via the transfer of a parcel of air or water (that was initially heated or cooled by conduction) from one point to another. This movement may result from density differences in the fluid (free convection) or from pressure exerted on it (forced convection). As with conduction, the total area of an object exposed to the moving fluid and the temperature difference between the surface of the object and the fluid are both important in determining total heat loss. To quantify heat exchange by convection, we also need a measure of the efficiency of heat gain or loss for a given orientation in the moving fluid. This measure is often expressed as a convection coefficient and is the equivalent of the thermal conductivity of the surrounding fluid. The rate of heat transfer by convection in air is, thus, influenced by the thickness of the boundary layer enveloping the object, and this in turn is influenced by the size and shape of the object, by its surface roughness, and by the wind speed. The equation describing heat transfer by convection is similar to that for conduction, with k taking on the value of thermal conductivity of air or water and d representing the thickness of the boundary layer across which the temperature difference between surface and surroundings is measured.

The third mode of energy exchange is by radiation. Unlike conduction and convection, radiant energy transfer does not involve molecular collisions and does not require any fluid medium. Radiation is simply the propagation of energy through space, as is the case with visible light energy, and can occur in a vacuum. The two properties of any object that determine the amount of energy it emits by radiation are its temperature (independent of ambient temperature) and its characteristic efficiency of radiant energy transfer, termed its "emissivity." Any object whose temperature is above absolute zero radiates energy in direct proportion to its temperature, with most natural objects emitting radiant energy at an efficiency of between 90 and 99% of the maximum possible for their temperatures. By measuring the temperature (T) of an object and knowing or estimating its emissivity (ε), we can, with reasonable accuracy, calculate its total energy loss via radiation using the equation

$$Q_r = \varepsilon \sigma T^4,$$

where σ is the Stefan-Boltzmann constant (5.67×10^{-8} $\text{W/m}^2 \, {}^\circ\text{K}^4$). This tells us that for most homeotherms heat loss by radiation remains fairly constant throughout the year since body temperature varies only slightly over time. Using the same equation, we can also easily calculate the amount of heat gained by an animal through the absorption of radiant energy emitted by its surroundings.

The final mode of energy transfer is by latent heat exchange. Latent heat is the amount of heat energy either tied up or liberated with phase changes of water. For example, the conversion of 1 g of ice at 0° C to water at the same temperature requires the addition of 335 joules of heat energy (i.e., 335 joules are tied up in the gram of water upon melting). This quantity is referred to as the latent heat of fusion. It is important to understand that this addition of energy does not alter the temperature any, but is required just to switch phase from solid to liquid. To convert 1 g of water at 20° C to vapor requires the addition of approximately 2450 joules of heat energy, the latent heat of vaporization. When the reverse process occurs, the same amount of heat is liberated. Latent heat exchanges in nature involve one of these phase changes, usually accompanied by a migration of water or vapor from source to sink.

The practical implications of latent heat transfer are many. Through this mechanism, for example, normal respiration in warm-bodied animals sets up something of a heat pump, with each exhaled breath removing heat by evaporation from the wet surfaces of the lung—heat that is robbed from the body as water changes phase from liquid to vapor. For those of us with active sweat glands in the wintertime, the benefits of reducing evaporative heat loss can be readily appreciated. The old trick of putting your feet in plastic bags before pulling on your socks and boots works well because the plastic barrier prevents the diffusion of vapor outward, thereby reducing the loss of latent heat that is tied up in vapor molecules (fig. 31). By keeping your socks dry at the same time, the rate of heat transfer by conduction is also reduced. The resulting condensation inside the bag may keep your feet wet with perspiration, but it also keeps body heat where you want it most. For the same reason, the common practice of installing vapor barriers behind the inside walls of our homes reduces latent heat loss and results in considerable fuel savings during the winter. Note that in both cases it is important that the vapor barrier be located on the warm side of the insulation. If

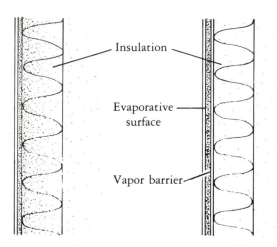

Fig. 31. How vapor barriers reduce heat loss. Diffusion of water vapor along a concentration or vapor pressure gradient (indicated by stippling) takes heat with it—about 2450 joules for every gram of water evaporated at the skin surface. A vapor barrier situated on the warm side of clothing (right) or on the inside of a building wall prevents this diffusion and thus reduces the loss of latent heat.

the vapor diffuses through the insulation only to condense against the cold plastic (liberating latent heat), then heat is lost quickly to the outside by conduction, regardless of our efforts.

WARM BODIES IN COLD ENVIRONMENTS

Physical versus physiological thermoregulation

We are now in a position to estimate the amount of metabolic energy that must be produced just to offset total heat loss during the winter and thereby maintain normal body temperature under cold conditions. To begin with, we will set up a simple word equation that says "Heat In = Heat Out." The "heat in" side of our equation represents energy produced through metabolism of food or fat plus any energy that may be absorbed from external sources (i.e., from sunlight). The "heat out" side of our equation represents the total loss by those processes that we have just discussed. This loss could now be calculated by combining separate equations for conduction, convection, radiation and latent heat exchange. This, in fact, would be the approach taken by the research ecologist who is interested in the relative contribution of each of these modes to total heat loss. However, for our purposes, certain simplifying assumptions can be made that render our task considerably easier and still provide valuable insight into the energetic advantages of the various behaviors we observe during the winter.

The first of these assumptions is that latent heat exchange is relatively small in most homeotherms during the winter, perhaps amounting to 10% or less of total energy loss, and thus can be ignored without serious error. This assumption is based in part on the observation in some small mammals of a 30% or greater reduction in body water turnover during winter.[1] Our second assumption is that the total heat loss from the outer surface of the skin, fur, or feathers by conduction, convection, and radiation is exactly equal to that which is conducted from the core of the animal to the outside surface. Because heat cannot be stored on the surface, no matter how heat loss from the outside is partitioned between conduction, convection, and radiation, the total must equal the amount conducted through the insulating layers from the core of the animal. This latter equality enables us to simplify our heat balance equation by using only the term for conduction to describe total heat loss from the animal (fig. 32). If we now assume that metabolic heat is the only substantial heat input (i.e., that the animal in winter does not absorb significant amounts of heat from its surroundings), then we can write another simple equation,

$$M = kA\frac{(T_b - T_a)}{d}$$

to represent the animal's energy budget, where M is the heat produced by metabolism. This now is the numerical equivalent of our "Heat In = Heat Out" equation.

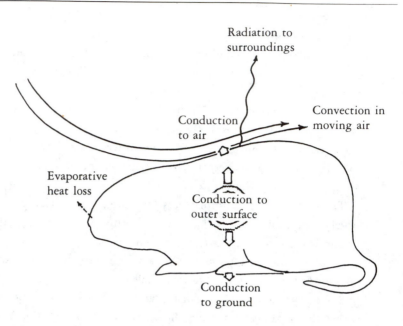

Fig. 32. *Heat loss from a small homeotherm. The total amount of heat lost from an animal's skin, fur, or feathers by conduction, convection, and radiation is equal to the amount of heat conducted from the animal's core to the outside of its body. Thus, heat loss in winter can be approximated by using a model for conduction alone, where it is necessary to know only the thickness and thermal conductivity of the animal's insulation, its surface area of exposure, and the temperature difference between the animal's core and its surroundings. Latent heat loss by evaporation is assumed to be negligible at this time of year.*

The advantage of this simple model is that we can use it to understand conceptually much of what physical temperature regulation is all about in the winter-active animal. The model tells us that there are several ways the animal can maintain a constant metabolic rate when faced with decreasing ambient temperatures. As air temperature drops further below body temperature (the quantity $t_b - t_a$ increases), metabolism can be held constant by either decreasing thermal conductivity (k), decreasing surface area of exposure (A), or some combination of the two. Also, the thickness of the insulating layer (d) can be increased behaviorally by erection of hairs

THE ENERGETIC ADVANTAGE OF HUDDLING

Because heat loss by conduction or convection varies in direct proportion to the amount of surface area exposed by the animal, the energy savings of huddling can be calculated easily. In the case of the small mammals pictured here, assume the following:

- Their effective surface area without huddling is 53.5 cm^2
- The thickness of their insulation (fat, skin, and fur) is .5 cm
- The thermal conductivity of their insulation is .004 W/cm/°K
- Their body core temperature is 37.5° C while the temperature of their surroundings is −20° C
- Their respiratory heat losses are negligible

Heat loss by conduction may now be calculated as follows:

$$Q_c = kA \frac{(T_b - T_d)}{d} = (.004)(53.5)(57.5/.5)$$

$$= 24.61 \text{ joules/second (or 24.61 watts).}$$

Now assume that by huddling each animal reduces its surface area of exposure by one-third. Their heat loss would then be reduced by an equal amount, as follows:

$$Q_c = (.004)(35.68)(57.5/.5) = 16.41 \text{ joules/second.}$$

In this same manner, the energetic advantage of nest building may be calculated by accounting for the effect of the nest on $T_b - T_a$, the body-to-air temperature difference, or by adding the thermal conductivity and thickness of the nest materials to the k and d terms, respectively.

or fluffing of feathers, as we so often see with roosting birds on a cold winter day. This has the effect of entrapping more air, thereby considerably increasing the insulative value of the fur or feathers. The surface area of exposure can be reduced, of course, by curling and retracting the extremities as much as possible or by huddling with other animals, and this too has a significant effect on the animal's energy balance. A one-third reduction in surface area, for example, by either of these means results in an equal reduction in heat loss (see box on "The Energetic Advantage of Huddling"). Finally, metabolism can be held constant through a reduction in the body-to-air temperature difference ($T_b - T_a$). This may be accomplished by elevating air temperature as much as possible through microclimate modification, such as nest building, or by exploiting the subnivean environment where temperatures are warmer. In some cases reduction of body temperature is also used to accomplish the same result. Some animals, particularly small birds, may undergo nightly torpor or controlled hypothermia, thereby decreasing their body temperature a few degrees and reducing heat flow accordingly.

Many small mammal species that are solitary during the summer become social during winter, constructing communal nests under the snowpack. Table 5 lists those noncolonial species for which there is presently some evidence of social aggregation during winter. The energetic advantages of such communal nesting are by now obvious. A beaver lodge in winter, occupied by two or more animals, remains well above freezing inside the central chamber, even when air temperatures outside are very low (fig. 33). Nests of taiga voles (*Microtus xanthognathus*) occupied by five to ten individuals have been found

Table 5. Noncolonial small mammals that show a tendency to congregate during winter

Clethrionomys gapperi—boreal redback vole

C. glareolus—redback mouse (European), bank vole

C. rutilus—tundra redback vole

Cryptotis parva—least shrew

Microtus californicus—California vole

M. montanus—mountain vole

M. ochrogaster—prairie vole

M. oeconomus—tundra vole

M. pennsylvanicus—meadow vole

M. pinetorum—pine vole

M. xanthognathus—yellow-cheeked vole, taiga vole

Peromyscus leucopus—white-footed mouse

P. maniculatus—deer mouse

Sources: From lists compiled by S. D. West and H. T. Dubun, "Behavioral Strategies of Small Mammals under Winter Conditions: Solitary or Social?" and D. M. Madison, "Group Nesting and Its Ecological and Evolutionary Significance in Overwintering Microtine Rodents," 267–74, both in *Winter Ecology of Small Mammals,* ed. J. F. Merritt, Carnegie Museum of Natural History Spec. Publ. 10 (Pittsburgh, 1984).

to remain between 7° and 12° warmer than ground temperatures and as much as 25° warmer than air temperatures above the snow. Furthermore, the nests are never completely vacated so that foraging animals always return to a warm nest.[2] In addition to this advantage of microclimate modification, the huddling individuals in the nest further reduce heat loss through effective reduction of surface area. It is not surprising, then, that metabolic rates have often been observed to be lower in individuals from communal nests than for animals tested singly (see numerous references cited in the work of D. M. Madison).[3] An added benefit of communal nesting may be a slight reduction in water loss as the higher relative humidity in the nest chamber reduces evaporation,[4] and this of course translates into reduced latent heat loss.

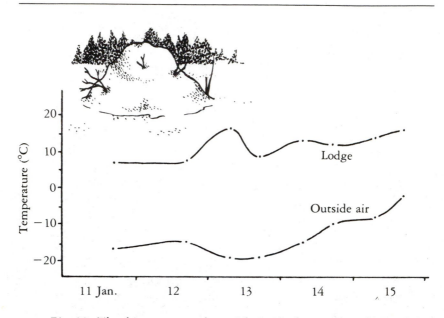

Fig. 33. The ultimate communal nest. The inside of a snow-covered beaver lodge may be as much as 35° warmer than the outside air in mid winter. This lodge was occupied by at least two adult beavers and was monitored by means of a thermistor probe inserted through the lodge wall and into the central chamber.

There are also some disadvantages to communal nesting, such as increased vulnerability to location by predators, a greater possibility of disease or parasite transmission, or more competition for food—but these may be small trade-offs compared to the problems of thermoregulation. With regard to the matter of food competition, communal nesters often show a reduction in body weight that may serve to reduce their total food requirements. Whereas this body size reduction would tend to increase cooling rate at the same time because of an increase in surface area relative to volume or mass, this disadvantage is far outweighed by the energetic advantage of huddling. In essence, communal nesters can well afford to give up body weight if the net result prevents increased competition for food.

The mechanisms that we have seen thus far for maintaining normal body temperature under cold conditions have involved only physical regulation of heat loss. By such behavioral modifications as curling, piloerection (the fluffing of fur or feathers), nest building, and huddling, animals are able to maintain body temperature over an impressive range of ambient temperature fluctuations without alteration of their metabolic rate. There are physical limits, of course, to the extent to which an animal can modify its environment or reduce its own thermal conductivity and surface area of exposure. In the face of still-decreasing ambient temperatures, the animal must at some point increase its metabolic rate in order to balance its heat loss. The temperature at which this becomes necessary is termed the "lower critical temperature" or LCT.

This lower critical temperature varies from one species to another and is seasonally adjusted in many animals, usually through changes in insulation thickness. The red fox, for example, reduces its LCT from 8° C in summer to − 13° C in winter. Likewise, the porcupine adjusts its LCT from 7° C to − 12° C. In both cases this is accomplished primarily through increased thickness of underfur.[5] In the harbor seal, parallel shifts in LCT from 22° C to 13° C in water and from 9° C to − 9° C in air[6] suggest that seasonal insulative adjustment is almost entirely the result of an increase in fat thickness. Smaller mammals, on the other hand, are somewhat limited in their ability to add fur or fat. The deer mouse increases the insulative value of its pellage in winter by nearly one-third,[7] but because its fur is still relatively short, the animal's LCT remains quite high. The same is true of the red squirrel, which has the same LCT in winter as in summer.[8]

Some small mammals lower their LCT slightly during winter by a seasonal increase in basal metabolic rate.[9] With a higher level of heat production to begin with, the animal can counter a small increase in heat loss before it becomes necessary to elevate metabolism in order to balance the heat budget. In general, however, the greater the ability of an animal to add fat and fur, the lower its winter LCT. At one extreme, the red squirrel, with a winter LCT of 20° C, is really not very well adapted to the cold and must rely on adequate nest insulation and high food consumption for its survival. At the other extreme, the arctic fox is superbly well equipped to survive extreme low temperatures, being able to remain comfortable at rest to temperatures as low as $-40°$ C without increasing its metabolic rate.[10]

The lower critical temperature, then, represents the point at which the animal has done all it can to maintain a constant metabolic rate through physical regulation of heat loss. Below this temperature, physiological thermoregulation takes over. The animal must now generate additional heat to match that which it is losing to its environment. Because many small mammals have lower critical temperatures well above $0°$ C, it is clear that metabolic heat production, or "thermogenesis" as it is called, is an important aspect of their winter ecology. Some seasonal adjustment of metabolism may be just as important to their overwintering success as the behavioral adaptations discussed so far.

An increased capacity for heat production without increasing muscular activity ("nonshivering thermogenesis") has been seen during acclimatization to cold in many small mammals. (In earlier discussions of plants and the winter environment the term "acclimation" was used to describe seasonal changes in cold tolerance. In the animal literature, however, this term is generally restricted in use to indicate short-term adjustments, usually under laboratory conditions, while the term "acclimatization" is reserved for long-term or seasonal adaptation.) This seasonal change is accompanied by, and is no doubt related to, an increase in brown fat deposits usually concentrated in the interscapular region of the animal, close to the vital organs. Brown fat is characterized by having a greater number of mitochondria and is thus capable of a higher rate of oxygen consumption and heat production than white fat. It also has a much higher density of nerve cells and blood veins than white fatty tissue. Brown fat is commonly found in abundance in young animals, including human infants, and was once thought to be only a precursor to white fat cells—"embryonal fat" it was called. But brown fat appears in adult animals, too, and is now known to play an important role in nonshivering thermogenesis. In fact, so effective is this tissue in generating heat that temperatures measured just under the skin over deposits of brown fat are sometimes higher than core temperatures.[11] This is most notable in the case of arousing hibernators, where brown fat acts as the heat generator for rewarming the animal.[12] An increase in brown fat shows up in red-backed voles beginning in early fall and continues to accumulate into winter, accompanied by a dramatic increase in the capacity of individuals for nonshivering thermogenesis.[13] These changes are cued by decreasing day length and amplified by low temperature; it is likely that the same mechanism is responsible for the increased capacity for thermogenesis seen during winter in white-footed mice, deer mice, Djungarian hamsters, prairie voles and shorttailed shrews.[14] Apparently the rise in heat production in brown fat upon demand is regulated by the hormone noradrenaline, which is secreted by the adrenal gland directly into the blood stream (fig. 34).

The capacity for producing heat through nonshivering thermogenesis also has its limits. When this mechanism is no longer sufficient to keep up with heat loss, then shivering takes over. For mammals, heat production by involuntary shivering is a last resort, but even this may be considered an adaptive mechanism where the capacity for shivering thermogenesis is subject to seasonal adjustment. In some northern mammals an increase in myoglobin (a protein similar to hemoglobin that binds oxygen) is seen in winter. The increased production of muscle myoglobin observed in red-backed voles during fall acclimatization parallels the seasonal increase in maximum total metabolic capacity in this species and is thought to facilitate oxygen transfer and storage, thus increasing the efficiency of shivering heat production by muscle tissues.[15]

With overwintering birds, thermoregulation takes on a very different pattern than that which we have seen in small mammals. While the same principles of energy exchange apply, the metabolism and behavior of birds is sufficiently different as to require a more physiological approach to regulation of body temperature, even at relatively warm air temperatures. Put another way, oppor-

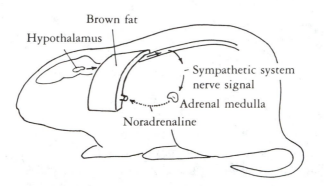

Fig. 34. Heat on demand from brown fat. Brown fat may be likened to a heating pad. As depicted in this schematic diagram, a nerve signal sent from the hypothalamus in the brain (the thermostat) to the adrenal medulla, a gland located next to the kidney, triggers the secretion of noradrenaline directly into the blood stream. It is this hormone that apparently stimulates rapid heat production in the mitochondria-rich brown fat tissue, which is usually concentrated in the interscapular region, close to the vital organs.

tunities for reduction of heat loss by physical means are much more limited among birds. A bird is restricted in its ability to reduce surface area of exposure by changes in posture, and behavioral mechanisms such as huddling are generally not used. Even the ability of the bird to add fat or feathers is limited by the mechanics of flight. Many year-round residents show both a seasonal and daytime increase in fat reserves during winter (see numerous references to this cited in the work of Nolan and Ketterson),[16] but generally birds add little more than enough fat to get through one long, cold night.[17] Nonmigratory dark-eyed juncos in Michigan, for example, are about 14% heavier during winter than their counterparts in Alabama as a result of increased fat reserves, but calculations based on overnight weight loss suggest that this extra fat gives the birds only a 16 to 24 hour advantage in fasting over their southern relatives.[18] Though this certainly may make the difference between success or failure in

Fig. 35. Fat and feathers aren't enough. Though fluffing feathers reduces heat loss, the insulation of small birds is not enough to keep them warm when air temperature drops. Lacking brown fat as well, the winter resident must shiver almost continuously to maintain normal body temperature. (Photo by Peter Marchand.)

the overwintering of northern birds, it is clear that the advantage of added fat is a metabolic one rather than a physical (insulative) one.

The addition of feathers in winter may be of less physical advantage than imagined, too. Birds in the subfamily *Cardueline*, represented by the goldfinches and redpolls among others, may increase total weight of feathers in winter by as much as 50% over their summer weight,[19] no doubt increasing their insulation some. Fluffing of feathers while roosting (fig. 35) might then reduce thermal conductivity of their plumage by another 30 to 50% as compared to daytime values.[20] But even so, studies of several species of northern birds indicate that their insulation is barely sufficient to allow maintenance of normal basal metabolic rates at air temperatures 10° below their body temperature.[21]

Birds, unlike the many small mammals that we have considered, also have limited opportunity to modify the temperature of their surroundings. Some, like the willow ptarmigan and several species of northern grouse, make use of the insulative value of snow by diving into the snowpack and tunneling a short distance to form a sheltered roost. Here they may spend anywhere from a few hours to three days at a time, benefitting from a much-reduced heat loss to their winter environment. If the temperature in the snow burrow remains 5° or so warmer than the surrounding snow and perhaps 25° warmer than the air above the snowpack, the roosting bird may profit by as much as a 45% reduction in energy loss.[22] The best most birds can do, though, is to seek the shelter of dense vegetation or the hollows in snow under shrubs, as redpolls often do, thereby reducing net radiant heat loss to their surroundings[23] (because some energy is gained by back-radiation from the cover around them, even though temperatures may be very low).

Instead of maintaining a constant metabolic rate over a relatively wide range of temperatures through physical thermoregulation, as was the case with small mammals, we see in birds a more gradual shift from physical to physiological adjustment at temperatures below 30° C.[24] This means that a curve of oxygen consumption in relation to ambient temperature is parabolic in shape, with no distinct break in the curve where physiological regulation takes over— that is, no well-defined LCT as we saw with small mammals (though there are LCT values occasionally reported for birds in the older literature). Another important difference in this metabolic curve in the case of birds is that the increase in oxygen consumption over its basal metabolic rate (the rate of metabolism while fasting and at rest) at lower temperatures does not come through nonshivering thermogenesis.

Birds lack brown fat and therefore cannot produce heat through nonshivering thermogenesis. Instead, they rely almost entirely on shivering to maintain body temperature. In fact, the available data indicate that even the larger birds like the crow and raven must shiver continuously during the winter when they are not generating heat through the muscle activity of flight.[25] An increased capacity for shivering thermogenesis during winter is seen in many birds and is apparently related to a seasonal increase in the lipid triglyceride (hence the fat increase that was noted earlier), which is also accompanied by a reduced rate of carbohydrate depletion.[26] Thus, it is not an increase in oxidative capacity of muscle tissue that is responsible for the increased thermogenesis in winter (the demands of flight already dictate a high oxidative capacity). Rather, it is an increase in the availability of fuel to keep the shivering muscle contractions going that enables the bird to produce heat longer at lower temperatures. Winter acclimatized goldfinches, when experimentally subjected to temperatures as low as $-70°$ C, can produce heat at a rate of 5.5 times their basal rate by shivering and can maintain normal body temperature for six to eight hours at temperatures below $-60°$ C.[27]

Chickadees show yet another degree of thermoregulation that involves lowering body temperature during inactive periods. This process, very similar to true torpor, reduces the rate of heat loss from the bird by minimizing the temperature gradient between body and air, and also reduces the amount of energy needed to maintain a stable body temperature. By controlling shivering through shorter and less frequent shivering outbreaks, body temperature gradually drops until a particular depth of hypothermia (varying seasonally) is reached. Shivering is then resumed with regular bursts, maintaining a closely regulated hypothermia. In chickadees this response is not dictated or necessitated by declining fat reserves, but is instead induced by decreasing temperatures and is utilized as a primary means of energy conservation. On a cold winter night the bird may allow its body temperature to drop $10°$ to $12°$, resulting in a savings of 20% of the energy normally required to maintain body temperature.[28]

If we assume that animals partition their use of metabolic energy in some prioritized way between the many life functions and activities that require energy (e.g., foraging, reproductive activity, growth, maintenance of body temperature), then thermoregulation must take the highest priority, especially during the winter. In the words of one expert, "If a [homeotherm] is not thermoregulating, then it is either hypo- or hyperthermic or is becoming so, and ultimately cannot perform its normal functions; therefore channeling energy to other functions would not be possible in any case."[29] Effective thermoregulation, then, is a prerequisite to the success of winter-active homeotherms. We have seen that maintaining a constant body temperature in the face of decreasing ambient temperatures is a multifaceted problem involving two levels of control, physical and physiological, with the latter incorporating both shivering and nonshivering thermogenesis. These controls are summarized schematically in figure 36 and tabulated for a number of species in the box "Balancing the Winter Heat Budget." In many cases these thermoregulatory mechanisms are so effectively used that the animal is able to remain comfortable under most winter conditions (to temperatures as low as $-70°$ C in the case of some small animals)[30] at an energy expenditure often no greater than that required during summer. Subnivean mammals may even have energy sufficient to support nonessential social interactions and, upon occasion, winter reproductive activity. However, animals that remain above the snowcover, often fully exposed to the icy winds of winter, have still other problems to contend with.

The problem with appendages

If a homeotherm didn't need its legs for mobility, it would be far better off without them in winter. The same may be true of all appendages that present a disproportionately large surface area to much colder surroundings. Imagine the problem of a beaver leaving the relative warmth of its dry, snow-covered lodge and slipping into the icy water for a trip to its food cache. As if the sudden loss of insulation weren't enough of a problem, the beaver's broad, flat tail

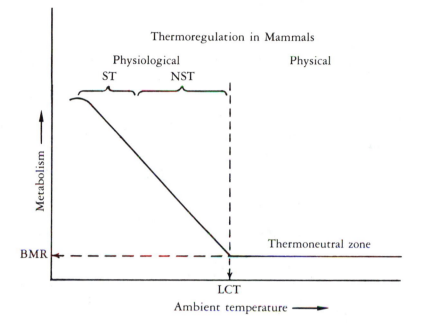

Fig. 36. *Physical versus physiological temperature regulation. Within the
thermoneutral zone, the animal's resting or basal metabolic rate (BMR) is
relatively constant, maintained through physical adjustment of surface area and
conductivity, such as by changing posture or by huddling. As air temperature
decreases and the limit of physical thermoregulation is surpassed, a point
designated as the lower critical temperature (LCT), metabolic rate and consequent
heat production increases. In most mammals this stepped-up heat production is
accomplished first through nonshivering thermogenesis (NST), which involves the
metabolism of brown fat, and then, as a last resort, through shivering or
increased muscle activity (ST). In birds the LCT is without clear definition.
Instead, there is a more gradual shift from physical to physiological
thermoregulation, with increased heat production below about 30° C accomplished
entirely by shivering.*

BALANCING THE WINTER HEAT BUDGET

A number of thermoregulatory strategies are employed by winter-active homeotherms to achieve the same result—maintenance of body temperature when the mercury drops. The table below is a compilation from the literature on the given species.

	Common shrew (Sorex araneus)	Redback vole (Clethrionomys gapperi)	Prairie vole (Microtus ochrogaster)	White-footed mouse (Peromyscus leucopus)	Black-capped chickadee (Parus atricapillus)	Common raven (Corvus corax)	Beaver (Castor canadensis)	Whitetail deer (Odocoileus virginianus)
Physical								
Increased insulation	X			X	X		X	X
Subnivean	X	X	X				X	
Communal nesting		X	X	X			X	
Physiological								
Torpor			No	X	X			
Adjusted BMR	X↑	X↑	X↑	X↑		No		X→
Increased NST capacity	X	X	X	X	No	No		?
Increased ST capacity		X			X	X		X
Weight adjustment	X↓	X↓	X↓	No				
Countercurrent heat exchange							X	X↓

X—known winter adaptation
Blank—not reported in the literature for a given species
No—known not to occur
?—likely, but evidence is inconclusive

in the water seems a perfectly designed radiator, sure to drain heat from the body faster than the animal could possibly generate it. And imagine the problem of a duck swimming in water only a degree or two above freezing, paddling with large webbed feet that seem created as much to dissipate heat from the body as to propel the bird; or the dilemma of a gull standing on an ice floe, circulating blood through its feet to keep those tissues from freezing and then returning the blood to its core for rewarming after giving up its heat to the ice.

The problem with appendages in each of these cases is that they must be supplied with oxygen and kept from freezing by circulating blood through them, but without allowing them to constantly drain heat from the body by conduction through an exposed surface that is disproportionately large relative to the volume or mass of the appendage. The solution to the problem is a physiological shunting of blood through a heat exchanger that intercepts the heat on its way out and maintains the extremity at a considerably lower temperature than the core, thereby substantially reducing heat loss. The heat exchanger in this case consists simply of veins and arteries in close proximity to each other and works as follows:

When air temperatures become low enough, superficial veins in the extremities constrict, increasing flow resistance and thereby shunting more blood through deeper veins lying in close proximity or in contact with the arteries. As cooler venous blood returning from the extremities flows parallel to the arteries carrying warm blood from the core, the temperature gradient between the two favors a transfer of heat from the arterial blood to the venous blood (fig. 37). In effect, a blood molecule traveling toward the core is exposed along the way to successively warmer molecules traveling in the opposite direction in the artery. With the veins and arteries in close contact, heat is easily conducted between them. The returning molecule continues to warm as it moves counter to a constant stream from the core, eventually reaching near-core temperature itself. The arterial blood, on the other hand, is continually giving up heat as it meets the steady stream of colder molecules

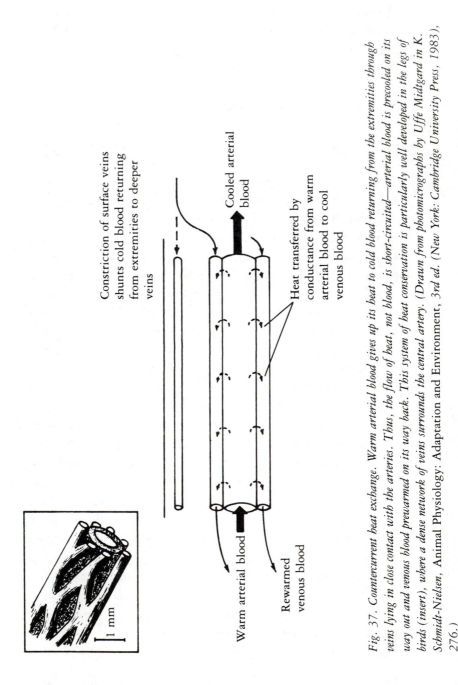

Constriction of surface veins shunts cold blood returning from extremities to deeper veins

Cooled arterial blood

Heat transferred by conductance from warm arterial blood to cool venous blood

Warm arterial blood

Rewarmed venous blood

1 mm

Fig. 37. Countercurrent heat exchange. Warm arterial blood gives up its heat to cold blood returning from the extremities through veins lying in close contact with the arteries. Thus, the flow of heat, not blood, is short-circuited—arterial blood is precooled on its way out and venous blood prewarmed on its way back. This system of heat conservation is particularly well developed in the legs of birds (insert), where a dense network of veins surrounds the central artery. (Drawn from photomicrographs by Uffe Midtgard in K. Schmidt-Nielsen, Animal Physiology: Adaptation and Environment, *3rd ed. (New York: Cambridge University Press, 1983), 276.)*

returning from the extremity. Thus, while the flow of arterial blood reaches the extremities, the flow of heat is short-circuited. Arterial blood is precooled on its way out and venous blood prewarmed on its way in.

This countercurrent heat exchange is especially well developed in some animals, with blood flowing through dense networks of branching and anastomosing arteries and veins in a system known as the rete mirabile ("miraculous net"). Blood circulating through the tail of the beaver passes through such a network and its effectiveness in exchanging heat is dramatic. In air at low temperature, heat loss from the tail may be reduced to less than 2% of the basal metabolic heat produced by the animal; whereas at higher temperatures, over 25% of the heat produced at resting metabolism may be dissipated through the tail.[31]

In whale flippers, each artery is surrounded by veins that precool the flipper and minimize heat loss to the cold water. The same arrangement is found in birds where heat exchange in the legs, particularly of aquatic or wading birds, is critical. An illustration of the artery and anastomosing veins surrounding it in the leg of a European rook is shown in figure 37. It is interesting to note that when the extremities are in imminent danger of freezing, increased blood supply and increased arterial pressure cause an enlargement in diameter of the artery, which then constricts the surrounding veins and forces more blood to return via superficial veins. Thus, an involuntary shunting of blood around the rete in effect delivers warmer blood to the extremities, since heat is not given up to the returning venous blood. When this occurs, the accompanying increase in heat loss to the environment is matched by an increase in metabolic heat production.[32]

BERGMANN'S AND GLOGER'S RULES: IS IT REALLY BETTER TO BE BIG AND WHITE?

These few considerations governing the loss of heat from warm bodies in a cold climate invite other questions regarding the possible adaptive significance of different body sizes and shapes. In a study of the energetics of woodrats and weasels at low temperatures, Brown and Lasiewski demonstrated that the spherical resting posture of the woodrat was far more efficient in reducing heat loss than was the oblong figure of the resting weasel.[33] Their conclusion from this study was that the weasel sacrifices some energetic efficiency as a result of being long and thin, in favor of increased hunting efficiency through its ability to get into small places. But this begs another question often raised in discussions of winter ecology: Is it better to be large or small in cold climates? The often-quoted Bergmann's Rule—that northern races of a species tend to be larger than southern races of the same species—implies that there may be some energetic advantage gained through an increased surface-area-to-volume ratio. The argument in support of this idea goes as follows:

Consider a cube, an imaginary little heat generator, 1 cm in each dimension. The surface area of the cube is 6 cm^2 while the volume is, of course, 1 cm^3. Hence, the surface-area-to-volume ratio is 6:1. Now double each dimension of this cube. The total surface area increases to 24 cm^2 while the new volume is 8 cm^3. The surface-area-to-volume ratio is now 3:1. Assuming that heat storage is related somehow to volume, and that heat loss is proportional to surface area, then this doubling of size would seem to favor heat conservation by reducing surface area relative to volume.

Admittedly, animals aren't cubes, but arguments favoring an energetic interpretation of Bergmann's Rule follow the same logic; i.e., that a larger animal has less total surface area per unit volume (equating volume with mass) and thereby benefits from a reduced cooling rate (see box on "The Size Trade-Off"). There are two problems with this interpretation, however. First, data on the body size of animals distributed over a wide latitudinal range show mixed results. In general, Bergmann's Rule seems to hold for some large mammals and nonmigratory birds. However, for small mammals, especially those that inhabit the subnivean environment, there is no evidence supporting an increase in body size northward. In fact, as already noted, individuals of some species show a seasonal reduction

THE SIZE TRADE-OFF

Some suggest that a larger animal benefits in winter from a more favorable surface-area-to-volume ratio. Others argue that more heat is lost from the greater surface area and, thus, it is absolute surface area, not relative area, that is important. Which argument is correct? For simplicity in calculation, consider heat loss from the two cubes illustrated here, where both are initially at the same core temperature, but where the dimensions of one are twice those of the other. Given the assumptions stated, it is readily apparent that the larger object loses more heat to its surroundings by conduction; but by virtue of its greater heat content, the larger cube also has more cooling "resistance" (i.e., cools at a slower rate).

Small cube		Large cube
125 cm^3	volume	1000 cm^3
.7 g/cm^3	density	.7 g/cm^3
87.5 g	mass	700 g
2.1 j/g	specific heat	2.1 j/g
150 cm^2	surface area (A)	600 cm^2
37° C	core temperature (T_b)	37° C
− 13° C	air temperature (T_a)	− 13° C
1 cm	thickness of insulation (d)	1 cm
.004 W/cm/°K	thermal conductivity (k)	.004 W/cm/°K
30 W	**heat loss by conduction***	**120 W**
183.7 joules	cooling "resistance"**	1470 joules
16.3° C	**core temperature drop after 100 seconds**	**8.2° C**

$$*Q_c = kA\frac{(T_b - T_a)}{d}$$

**Heat loss required to lower T_b by 1° (specific heat x mass)

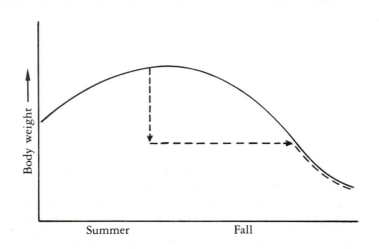

Fig. 38. Weight reduction in preparation for winter. Laboratory voles on a restricted diet maintained an artificially low body weight through the summer (dashed line), but still matched the weight loss of their field cohorts (solid line) in the fall, even when given excess food. (Data from Bruce Wunder, personal communication.)

in body size as winter approaches. This is apparently genetically controlled, perhaps cued by decreasing day length.[34] Bruce Wunder at Colorado State University placed voles in the laboratory on a restricted diet to artificially reduce and hold their body weight until their field cohorts' naturally declining weight reached the same level.[35] The voles on the restricted diet were then given surplus food, but instead of gaining weight they keyed into the declining weight curve of their cohorts and commenced losing weight in the same manner (fig. 38). Similarly, free-living voles born in the fall, after only a brief period of growth, key into the same declining weight curve of their earlier spring- and summer-born cohorts, which grow initially to a much larger size.[36]

The second problem with an energetic interpretation of Bergmann's Rule, as argued by Brian McNab, is that animals don't necessarily know anything about their per-gram efficiency.[37] Their only concern is with total food requirements, and the larger animal needs more food, a decided disadvantage during long winters in the North. As an alternate explanation where Bergmann's Rule does seem to hold, McNab suggests that the increase in body size northward is the result of a release from competition. He supports his argument with data showing that the body size of the shorttail weasel, among other animals, increases northward only after it extends beyond the northern range limits of the longtail weasel, its closest competitor (fig. 39). This appears to be the case with other close competitors, too, as for example with the fisher and marten. In western Canada the marten shows an increase in size only north of about 62° N latitude, the range limit of the larger fisher; but in Labrador, where the fisher drops out at 52° N latitude, the marten shows a corresponding size increase immediately.[38]

It is possible, then, that size differentiation is not a primary adaptation to optimize heat balance, but is instead related to prey size selection amd competition among similar species for limited food resources. In the case of voles that lose weight at the beginning of the winter, the reduced food requirement appears to outweigh the disadvantage of an increased cooling rate from the smaller body. This is especially true with the communal nester that can afford to lose weight, since huddling offsets the energetic disadvantage of smaller size.

The issue of coloration as it relates to energetics is a bit more complex. It is often suggested that black coloration, because it absorbs more sunlight, also radiates more heat and would, therefore, be disadvantageous to a homeotherm in cold climates. A corollary of this argument, that white radiates less heat, is sometimes suggested as the ecological basis for Gloger's Rule, which tells us that pigmentation in animals tends to be reduced towards the poles, with northern races of animals generally being lighter in coloration than their southern counterparts.

That a black object should radiate heat energy with greater efficiency than a white one would be true only if the black object were heated substantially more as a result of greater absorption of sunlight (assuming the two have equal emissivities). Early discussions

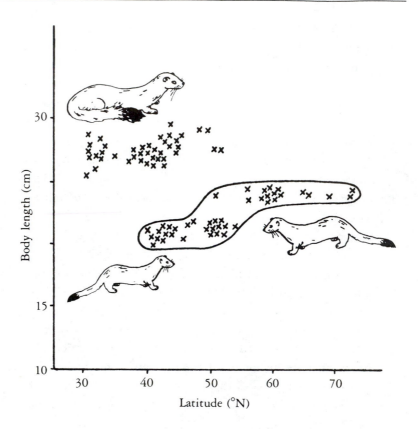

Fig. 39. Body length of male weasels, Mustela frenata *(top) and* M. erminea
*(bottom), related to latitude. (Dàta from B. K. McNab, "On the Ecological
Significance of Bergmann's Rule,"* Ecology, *52 (1971): 845–54.)*

noted that radiant heat loss from any object is largely a function of
its temperature. However, the homeotherm, by definition, maintains a constant body temperature regardless of whether it is white
or black. The animal with black coloration may warm more by absorbing solar radiation, but it compensates for this heat load by
regulating other modes of heat production or dissipation so as to
maintain a constant core temperature. Therefore, it loses no more
heat by radiation than does the white animal. This balance between

Fig. 40. The energetic advantage of being black. Zebra finches dyed black metabolize at a considerably lower rate in daylight than do all-white birds because of their greater absorption of sunlight. The use of external sources of heat to balance an animal's energy budget reduces the requirements for thermogenesis. (Data from W. J. Hamilton, III, and F. Heppner, "Radiant Solar Energy and the Function of Black Homeotherm Pigmentation: An Hypothesis," Science, 155 (1967): 196–7.)

heat absorption and metabolism has been demonstrated under controlled conditions by comparing the metabolic rates of some all-white zebra finches with that of the same finches dyed black (fig. 40). With the "sun" turned on in the experimental chamber, the all-black birds metabolized at a considerably lower rate than the white birds, thus enjoying some energetic advantage from the absorption of light.[39]

This being the case, one wonders then why it isn't of greater advantage, from a purely energetic point of view, to be black in the North rather than white. We think immediately of the importance of cryptic coloration, but what is the advantage to the arctic fox, a true scavenger with little worry about predator avoidance, of being white rather than black? It is interesting to note that in the Pribilof Islands, and in many coastal areas of Alaska and Canada, it is the blue phase of the arctic fox that is most common. If it is important for some reason that the arctic fox blend in with its surroundings, then why do these animals complicate their lives by remaining slate-blue in color throughout the winter? And what is the advantage to the shorttail and least weasels of turning white in winter, when they spend most of their time hunting small mammals under the snow, where visual clues are relatively unimportant to predator or prey? If cryptic coloration were important to the hunting success of preda-tory animals, then one might expect that the red fox or the weasel's larger relative, the fisher, who hunt above the snow, would turn white in winter, instead. And why does the arctic hare retain its conspicuously white coat all summer on Ellesmere and Baffin Islands and in Greenland?

In an effort to answer some of these questions concerning adaptive coloration in homeotherms, W. J. Hamilton, III, studied the phys-ical and behavioral characteristics of 27 species of all-white and 39 species of all-black birds in western North America.[40] Among his findings he noted that, in general, the insulation of the all-white birds was far superior to that of the black birds, and that this dif-ference was reflected in their roosting characteristics. All of the white birds spent the night in the open, whereas, with only few exceptions, the all-black birds roosted under cover, where they de-rived some benefit from back-radiation of heat from their surround-ings. Hamilton suggested from these data that in birds, at least, white coloration serves to minimize heat exchange with the environ-ment. They reduced loss through increased insulation while at the same time reducing heat gain through increased reflectance, which helps to prevent overheating when the animal is active during the day.

Is the increased insulation of the all-white birds in Hamilton's study merely coincidental, or is there a relationship between color and insulation? Whiteness in most terrestrial animals is due to the absence of the granular pigment melanin. When lacking melanin, the hair of mammals is highly vacuolate (hollow), with the result that light is scattered and reflected rather than being absorbed, and thus it appears white. The same phenomenon is responsible for the reflectance of white feathers. The fact that the white hair or feather barbule is hollow has another important implication. Leaving air spaces in place of pigment granules surely reduces the thermal conductivity of the pelage or plumage. Thus, there may indeed be a physical reason for the better insulation of the white birds and for the superiority of the white arctic fox's insulation over other animals with fur of equal thickness.[41]

Unfortunately, there are no conclusive data in the literature to help resolve our questions here. Even the stimulus for color change in northern animals gives little clue as to the real advantage of white coloration. In some species, temperature appears to control the seasonal change in color, but in others, even within the same genus, photoperiod may be of overriding influence. Experimentally, both short days and low temperatures, by themselves, have been found to induce a change from summer to winter coloration, while exposure during winter to long days for some animals or high temperatures with others can reverse the process. In either case, under natural conditions both temperature and photoperiod would serve to time the molt so as to coincide with annual changes in snowcover. Thus, we can't rule out the possibility that snowcover is still the primary selective pressure that led to the evolution of seasonal color change. Perhaps increased insulation is a fortuitous secondary advantage. We can only agree that there is more to white coloration in winter than meets the eye, and that the possible energetic advantage of having white fur or feathers remains an interesting question in the winter ecology of birds and mammals.

5 LIFE UNDER ICE

Our consideration of winter ecology began with the arrival of a continuous
snowcover on the land, but in the aquatic world it is a different
beginning—one which takes us further back in time to the first hint
of the coming season. As the days of autumn begin to cool and the
morning dew turns to frost, changes are set into motion in the
aquatic ecosystem. One event gaining momentum rapidly at this
time will have a profound influence on all life under the coming ice
cover. It is the process of overturn—the mixing of surface and bottom waters that redistributes resources in the aquatic environment.

To understand how this overturn works and why it is so important to organisms under the ice, it is first necessary to understand
something of the unusual relationship between the temperature of
water and its density. As warm water cools, it, like most liquids,
becomes more dense. Unlike other substances, though, water
reaches its maximum density at 4° C. Thereafter as it continues to
cool, water actually becomes lighter, undergoing its most pronounced change when it is finally converted to ice. This tempera-
ture/density relationship (fig. 41) is a fortuitous quirk, for if water
behaved like most other liquids and continued to increase in density
as the temperature dropped, then ice would form first on the bottom
of the lake, eventually freezing right to the top!

It is in large part this unusual temperature/density relationship
that drives the process of overturn in the lake environment. During
the summer, a temperature stratification usually exists in deeper
lakes as surface waters are warmed and float above the colder, denser
bottom waters. Wind currents may stir this surface layer some, but
mixing to any depth is resisted by the density differences that ac-
company the temperature stratification. Thus, throughout the summer the oxygen-rich waters remain at the surface. However, when
air temperature begins to drop in the fall of the year, the stratification that prevents mixing starts to break down. As the surface of
the lake cools, the slightly denser water gradually sinks, mixing
now with water of lower strata. But the cooler water can sink only
to the depth at which it encounters water of the same temperature
and density. Colder, heavier water below prevents further subsid-

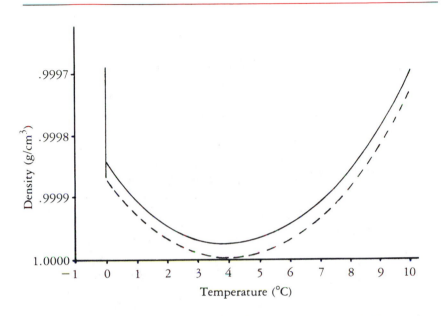

Fig. 41. Temperature-density relationship for ordinary water (solid line) and pure water free of air (dashed line). The density of ice at 0° C is .917 g/cm³.

ence. As the season progresses, however, the continually cooling surface water sinks to greater and greater depths, until finally the whole water column reaches the same temperature from top to bottom. Once this occurs, there are no longer any density differences to resist mixing, and even a gentle wind stirring the surface water can now generate vertical currents that reach all the way to the lake bottom (fig. 42*a–c*). This top-to-bottom turnover of the water in the fall is an event of profound impact, for it results in the complete circulation of nutrients and oxygen throughout the lake—an important precedent for the hard times ahead. The lake is saturated with oxygen again, and the colder water holds more now than in the summer (the concentration of oxygen saturating water at 5° C is 12.37 ppm compared to 8.84 ppm at 20° C), but this time it is a finite supply and it may be a long winter under the coming ice cover.

As winter draws nearer, the temperature of the entire water col-

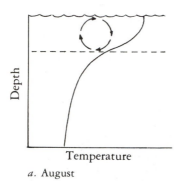

a. August

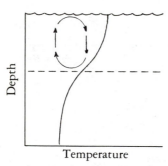

b. September

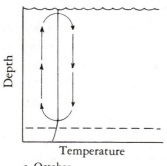

c. October

*Fig. 42. Overturn of water as temperature drops. (*a*) Warming of surface water results in a steep temperature gradient and stratification of a lake due to differences in the density of water. Mixing of water by wind currents is limited to a relatively shallow surface layer. (*b*) Cooling surface water subsides to a depth of equal temperature, reducing temperature and density stratification in the upper part of the water column and allowing deeper mixing. (*c*) Continued cooling and subsidence has eliminated temperature and density stratification so that mixing currents generated by surface winds now reach to the bottom of the lake. This top-to-bottom circulation is referred to as "overturn" and is important in redistributing oxygen and nutrients before ice cover seals the lake.*

umn moves toward the point of maximum density (4° C). With continued cooling, water no longer sinks but becomes increasingly lighter and it stays on top. A reverse stratification develops. The 4° C water at the greatest depth cannot be displaced, so in a lake a few meters deep the bottom water is likely to remain at that temperature throughout the winter. Wind may continue to stir and oxygenate surface waters before freezing, but finally, on a calm night, an ice skim forms and brings an end to the mixing. This is the beginning of winter in the aquatic world. The thickening ice cover is like a lid on the ecosystem. The exchange of gasses between the lake and atmosphere is effectively shut off now, and what life-supporting oxygen is dissolved in the water is all that the aquatic fauna will have for the duration of the winter.

With an ice cover on the lake, the entire temperature range for life is now narrowed to 4° (perhaps 5° if we move into the bottom mud). The aquatic invertebrates and fish that remain active in this environment are all "cold-blooded" poikilotherms, unable to control body temperature independently of their surroundings, which means that they must function at body temperatures within the same narrow range as the water. Because chemical reaction rates in the body are more or less directly linked to temperature, it would not be surprising, then, to find the metabolic rates of these organisms greatly slowed. In fact, as a general rule, for every 10° change in temperature, the metabolic rate of the poikilotherm increases or decreases by a factor of two or slightly more. This means that if a fish were acclimated to a temperature of 25° C and we abruptly put it into water at 5° C, its body functions would slow to only one-fourth their normal rate (one-half for the first 10° and half again for the next 10°). Thus, some aquatic organisms become very inactive at the low temperatures of winter, remaining in a state of "semihibernation" in which they are slow breathing and lethargic, but can still respond to physical stimuli. Inactive fish like bluegills and the many species of zooplankton that remain suspended in the water column are aided by the higher density of the cold water that in-

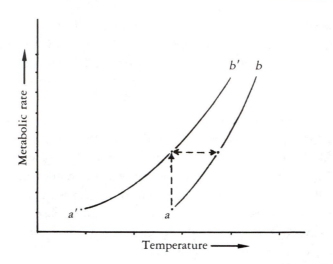

Fig. 43. *Upward adjusted metabolic rate in aquatic organisms during winter. The relationship between temperature and metabolic rate in a warm-acclimatized poikilotherm is given by the curve* ab. *However, in the process of cold acclimatization this relationship shifts to the left (curve* a'b'*) such that the organism in winter may maintain the same rate of metabolism that was formerly possible only at a higher temperature (horizontal dashed line). Comparing metabolic rates for any given temperature shows an upward adjustment in the cold-acclimatized organism (vertical dashed line).*

creases their buoyancy and minimizes the amount of energy needed to stay there.

However, this metabolic "rule" was not made for all. Many aquatic organisms show a remarkable ability to adjust their metabolic rate over time as they acclimatize to cold water. As water temperature cools gradually with the approach of winter, the temperature tolerances of these organisms shifts, too. The winter-acclimatized organism in some cases can maintain the same level of activity at a lower temperature as was formerly possible only at a much higher temperature. A salamander, for example, that has been acclimatized to a temperature of 15° C might show a metabolic response to rapid temperature changes as indicated by curve *ab* in figure 43. In this case, its metabolic rate is considerably reduced when exposed to colder water. However, if held at the lower temperature for a period of time, its metabolism gradually increases, a response often referred to as an "upward adjusted" metabolic rate. In effect, the whole curve shifts to the left (curve $a'b'$ in figure 43), with the salamander soon metabolizing at the same rate as it did previously at 15° C. This adjustment has been observed in red-spotted newts that remained active throughout winter at the bottom of a shallow pond, where a spring kept a small area of water ice free. A similar response has also been seen to varying degrees in fishes, crayfish, and sand crabs.[1]

The ability of a poikilotherm to shift to higher metabolic rates at lower temperatures apparently involves a change in enzymatic reaction rates, which can come about in one of two ways. An enzyme that serves as a catalyst for a specific metabolic reaction may exist in two or more different forms called isozymes. Isozymes are variations of the same basic molecule, but they often exhibit different temperature sensitivities so that one or another controls a reaction at a given temperature. Thus, adjustment to decreasing temperatures may come through a change in proportions of different isozymes. If, for example, isozyme "x" exhibits maximum sensitivity at a temperature of 10° C and isozyme "y" shows a maximum response at 5° C, then an increase in the ratio of y to x would result in greater activity at the lower temperature. In figure 43, idealized for a number of organisms, the convergence of the two curves toward their upper end suggests such a change in isozyme ratio.

Another means by which metabolic adjustment can be achieved is simply through a change in enzyme concentration—the production of more enzyme to catalyze more reaction. If, in our winter-acclimatized animal, metabolic rates were increased by an equal amount at all temperatures (i.e., curve $a'b'$ in figure 43 was parallel to curve ab), the implication would be that an increase in total number of enzyme molecules had occurred rather than a change in isozyme ratio.

Whichever way it is accomplished, an upward adjusted metabolic rate is clearly of adaptive value in allowing many organisms to function throughout the long winter under ice. But low-temperature acclimatization is not the final answer to winter survival, for the organism that remains active faces still other problems. In some lakes it is the progressive depletion of oxygen in the water, rather than low temperature, that creates the greatest stress of winter.

All aquatic organisms, including the hibernating reptiles and amphibians in the bottom mud, utilize oxygen throughout the winter. With supply limited now, a gradual deficit accumulates through respiratory consumption of oxygen. The mud-dwelling decomposers are the most numerous and heaviest users of oxygen, so it is usual to see the oxygen depleted first near the bottom as it is removed by these organisms. Oxygen then diffuses downward through the water column toward areas of lower concentration, but only very slowly, giving rise to an oxygen profile or concentration gradient such as that shown in figure 44. It is not uncommon by mid winter to see a near absence of oxygen in the bottom waters. Ice fishermen are well aware of the manner in which many fish species rise to higher strata in the lake as oxygen depletion progresses. As winter wears on, only slow internal currents—perhaps some slight inflow-outflow movement of water—will help redistribute oxygen within the lake.

If the shortage of oxygen becomes severe enough, the less tolerant organisms may die. Fish kills from oxygen starvation are not uncommon in shallow, productive ponds during unusually long winters. When the concentration of dissolved oxygen reaches critical levels, the more resistant organisms may switch to anaerobic respiration in which they derive energy from the breakdown of carbohydrate (usually glycogen) to lactic acid—a process known as glycolysis. This oxygen-saving metabolism is utilized also by hibernating turtles (and perhaps by the amphibians as well) to supply their energy needs throughout winter.[2] Glycolysis yields only about 8% of the energy that would be available from the complete oxidation of glycogen in aerobic respiration. In some cases, the accumulated lactic acid can still be used as a substrate for later oxidation reactions, thus eventually giving back all of the energy of the original carbohydrate.[3] In

Temperature (°C)   Dissolved oxygen (ppm)

Fig. 44. Temperature and dissolved oxygen profiles typical of mid winter in a moderately productive temperate lake 5 m deep. Lakes with lower productivity (fewer oxygen-consuming organisms) will have a more vertical profile, with a higher oxygen concentration in the lower strata.

some fish that are especially tolerant of oxygen-deficient conditions, the accumulating lactic acid may be converted to ethanol, which then diffuses readily from the gills. This wastes a potential source of energy, but reduces the acid-balance problems caused by excessive lactic acid.[4]

Aquatic plants might be singled out as major contributors to winter oxygen stress but for the fact that some species appear adapted to photosynthesizing at the very low light levels and low water temperatures typical under ice. This, of course, gives back some of the oxygen used in respiration. Many species of phytoplankton—minute suspended plants including motile algae—remain photosynthetically active throughout winter, shifting in their vertical distribution as light levels change with snowcover conditions on the ice[5] and contributing substantially to the fixation of carbon and release of oxygen in the water. Elodea (*Elodea canadensis*), a common submerged aquatic plant of northern lakes, is known also to increase in biomass under ice,[6] and it appears that much of this wintertime growth is supported by active photosynthesis. Plants that have been gathered through holes in the ice, placed in closed glass containers, and then returned to the pond bottom to "incubate" under normal winter conditions have shown considerable amounts of oxygen evolution as a result of photosynthesis.[7] In fact, in January, under the lowest light levels occurring beneath the ice, these plants produced enough oxygen during daylight hours to pay back 40 to 60% of the oxygen they consumed in respiration over a 24-hour period. At times during January, oxygen production reached 50% of the gross daytime release seen by this species in late summer. This is enough to cause a significant increase, during daylight hours, in the dissolved oxygen concentration of water above a bed of submerged aquatics. Figure 45 shows the daily change in ambient dissolved oxygen attributable solely to vascular plants photosynthesizing under remarkably low light filtered through 47 cm of snow and ice and 1 m of water at a temperture of only 1.2° C. While there is still a net reduction of dissolved oxygen at the end of 24 hours, the daytime pulse is significant. Later in the season,

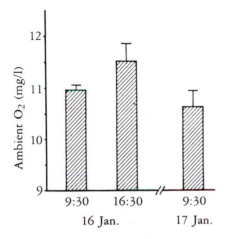

Fig. 45. *Daily change in dissolved oxygen concentration in the littoral zone of a shallow lake as a result of photosynthesis by vascular plants alone. The water temperature on January 16 was 1.2° C and maximum irradiance was .61 μE/m/s (.15% of incident above a thickness of 47 cm of snow and ice cover). (Data from P. J. Marchand, "Oxygen Evolution by* Elodea canadensis *under Snow and Ice Cover: A Case for Winter Photosynthesis in Subnivean Vascular Plants,"* Aquilo Ser. Botanica, 23 *(1985).)*

when days are getting longer and light levels under the ice increasing, oxygen evolution increases substantially, with payback exceeding 75% of respiratory consumption. This raises interesting questions about the possibility of daytime migration of other organisms into the shallow littoral zone in response to oxygen production by aquatics.

This photosynthetic ability does not exonerate plants from their part in the progressive development of oxygen stress in the aquatic environment under ice, but without it things might be worse. These plants highlight the remarkable ability of aquatic organisms of all kinds to adapt to the unique stresses of this season. In the end, though, it is a game of endurance, for winters under the ice are very long. A January thaw may bring temporary respite to the land, but seldom does it break the seal on the aquatic ecosystem. Spring melt, too, is slow to release the lake, and even then it is only with the occurrence of spring turnover—a repeat of that fall process that again stirs new life into the deep waters—that winter finally ends in the aquatic world.

6 PLANT-ANIMAL INTERACTION: FOOD FOR THOUGHT

Earlier in our discussion of plants and the winter environment, we talked about the mechanical forces of winter—the problems of snow loading, drifting, and ice blast as they affect trees and shrubs exposed above the snowpack. We considered how these forces mechanically shape trees at high elevations, how they affect physiological processes like water relations and carbohydrate balance, and how they influence demographic processes like wave-mortality and the development of ribbon forests in some mountain areas of the western United States. But we did not talk about the ecological impact of concentrated winter browsing by herbivores on woody plants. In areas of the northern coniferous forest where local concentrations of hares or moose are high, or in areas where deer yard for the winter, the browsing of woody plants above the snowpack may constitute a force of major significance in the evolution of plant survival strategies. In some situations, such as the early development of forest stands on river floodplains in Alaska, browsing of willows during winter by snowshoe hares can be one of the most important factors limiting growth of the plant and influencing stand succession. Likewise, in Scandinavia the grazing pressure on willows and birches during winter by the mountain hare is often reported to have a major impact on the development of the plant community. This pressure obviously poses a serious problem to the plant, but it may have a long-term impact on the grazers as well.

Plants that survive heavy winter browsing often show distinct changes in growth form as internal allocations of nutrients and carbohydrates are shifted about. For example, birch saplings clipped by browsing moose may respond by producing heavier shoots with larger, more chlorophyll-rich leaves that are retained longer by the plant. In addition, the resting buds on mature branches often produce "long shoots"—shoots with elongated internodes between leaves.[1] Willow responds to heavy browsing by producing numerous stump sprouts of juvenile form, consisting mostly of long shoots without lateral branches.[2] This "compensatory growth" is typical of the response of many species to herbivory and can itself alter the microclimate of the plant community, affecting the future photosynthetic efficiency of the plant as well as influencing later availability of browse to the grazers. Compensatory growth, though, is not the plant's only answer to winter browsing pressure.

Several studies of feeding behavior in hares during winter indicate a strong selectivity in browsing preference. The green foliage of conifer saplings and evergreen shrubs such as labrador tea are usually passed up in favor of the twigs of deciduous trees and shrubs. Within this preferred group of plants, hares show a consistent preference for mature growth rather than juvenile shoots, even though the latter may be more plentiful and more accessible during the winter. Evidence suggests that this choice is not based upon nutrient content or protein level of the preferred forage, but rather is related to the presence of feeding deterrents in the juvenile shoots and in the buds and catkins of the mature plant. Many plant species, it turns out, have evolved strong chemical defenses in response to grazing pressure—unpalatable or toxic compounds that serve as deterrents to browsing. Plants that typically inhabit high-resource environments and have rapid growth rates, such that they soon grow out of the reach of grazers, may produce antiherbivory compounds only in their juvenile parts. Other plants inhabiting low-resource environments and having slow growth rates with less potential for compensatory growth, typically maintain chemical defenses throughout their life cycle.[3]

In Alaskan paper birch, the chemical compound that serves as the deterrent appears to be papyriferic acid, which is present in juvenile stems at concentrations 25 times greater than in mature stems.[4] So great is the defense level of this species in boreal forest regions of Alaska, where browsing pressure is especially high, that resin droplets actually accumulate on the surface of juvenile twigs (Fig. 46). The effectiveness of papyriferic acid as a feeding deterrent has been demonstrated by treating oatmeal with purified extracts from the birch resins and offering it, along with untreated oatmeal, to captive snowshoe hares. In these experiments the hares consumed only one-fourth as much of the treated oatmeal as the untreated feed, and when offered only juvenile birch twigs with naturally high con-

Fig. 46. Mature (left) and juvenile twigs of Alaskan paper birch. Chemical defense of young shoots against browsers like the snowshoe hare is so pronounced in this species that resin droplets of the deterrent chemical actually accumulate on the surface of the juvenile twig.

centrations of the acid, the captive hares eventually stopped eating entirely.[5]

Avoidance of juvenile willow branches by hares may be related to higher lignin content and lower digestibility of nitrogen[6] or to increased concentrations of the compound phenolic glycoside.[7] When free-ranging mountain hares in Finland were offered bundles of different-aged willow and aspen twigs along with oats that were either untreated or treated with extracts of phenolic glycoside from willow bark, the results were similar to that seen with the birch. Mountain hares showed a strong preference for mature shoots of both species and avoided the treated oats in favor of the untreated, apparently able to discriminate between the food items by scent alone.[8]

Some researchers suggest that preferential browsing of mature willow shoots by snowshoe hares may have an important feedback effect on the hare population itself. In boreal forest regions of North America and the Soviet Union, hare population cycles have a regular periodicity of 8 to 11 years, which is thought to be the result of a time delay in vegetation recovery following overbrowsing during winter. As we have seen here, when some woody plant species are browsed heavily, they respond by producing juvenile shoots that are

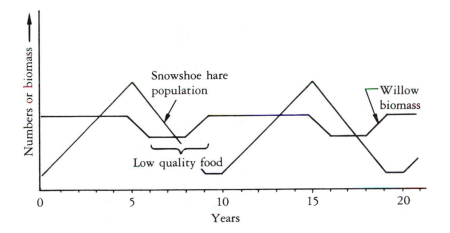

Fig. 47. Time lag in recovery of hare population keyed to plant defenses. Willows and birches respond to overbrowsing by producing juvenile shoots with strong chemical defense compounds that deter further browsing. These chemical defenses are maintained for two to three years, thus creating a food shortage that extends beyond the recovery of plant biomass following the hare population decline. This delay in recovery of food quality probably contributes to the lengthened periodicity of snowshoe hare population cycles. (Adapted from L. B. Keith, "Role of Food in Hare Population Cycles," Oikos, 40 (1983): 385–95.)

of very low food value and that typically maintain their chemical defenses for a period of two to three years. In effect, as browsing pressure increases during the growth phase of the hare population cycle, the resistance of plants to further browsing also increases. It appears now that this may be the mechanism that limits hare population growth and initiates the declining phase of the population cycle.[9] This evolved chemical response of plants to browsing effectively extends the food shortage beyond that induced by overbrowsing alone and, thus, provides a time lag in recovery of food resources (fig. 47). Meanwhile, declining hare reproduction (induced by a scarcity of high-nutrient food resources) and high winter mortality, perhaps coupled with increased predation pressure following the population peak, would continue to drive the hare population down.

Fluctuations in microtine rodent populations in Finland have also been linked to plant-animal interactions, in this case involving changes in both plant nutritive quality and chemical responses to grazing.[10] Arctic and subarctic plants themselves exhibit cyclic patterns in flowering and fruit production due to the general scarcity of resources in that environment. One growing season is rarely long enough to allow accumulation of the necessary carbohydrate and nutrient reserves for flowering, so plants accumulate reserves for several summers until some threshold is reached and flowering commences. It follows, then, that plants in flowering condition are more vigorous and of potentially greater food value. During the winter prior to flowering, rootstocks or rhizomes of perennial herbs are high in carbohydrate reserves and provide an important source of energy to subnivean mammals. During the winter following flowering, high energy seeds or fruits are abundant. By the second winter after flowering, the nutritive value of the plant is low again. Cyclic patterns of plant growth are not always synchronous among species in northern Finland, but the periodicity of flowering and seed production is fairly regular at three- to four-year intervals and corresponds closely with the rise in small mammal populations, often preceding the peak by one year (fig. 48).[11]

Following heavy grazing by small mammals, these herbaceous plants may also increase production of defense compounds, thereby having additional feedback on the herbivore populations. Because synthesis of defense compounds competes with other growth functions, such as leaf production, for nutrients and metabolites, an increase in production of antiherbivory compounds in areas where nutrient resources are scarce is likely to slow the rate of plant regrowth.

Fig. 48. Cyclic patterns of plant production and rodent populations in northern Finland (From S. Eurola, H. Kyllonen, and K. Laine, "Plant Production and Its Relation to Climatic Conditions and Small Rodent Density in Kilpisjarvi Region (69°05′N, 20°40′E), Finnish Lapland," in Winter Ecology of Small Mammals, *ed. J. F. Merritt, Carnegie Museum of Natural History Spec. Publ. 10 (Pittsburgh, 1984), 121–30.)*

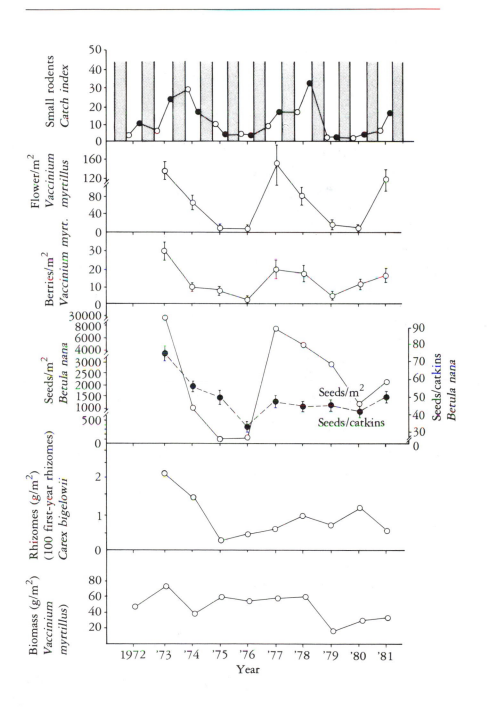

Recent observations in Finland suggest that induced chemical response by herbaceous plants rises and falls with changes in grazing pressures and, thus, may have the same time delay effect on animal population recovery as seen in the snowshoe hare/willow interaction.[12] Superimposed on the interaction here is the effect of increased predation following the peak in rodent numbers as well as possible parasite and disease interactions which also affect the rate of population recovery.[13] These interactions are summarized in figure 49. It is interesting to note that arctic hare populations in northern Scandinavia follow the same three- to four-year cycle of microtine rodents rather than the 10-year cycle of snowshoe hares in North America and the Soviet Union. This is apparently because predators switch over to the hares during the decline of rodent populations.[14]

We should not overlook the possibility, too, that plants might directly mediate the occasional winter breeding that is documented for some subnivean mammals. Increasing evidence suggests that the stimulus for breeding in small mammals may come from plant compounds ingested by the animal. Compounds that induce reproductive activity in captive rodents have been isolated[15] and include gibberellic acid, a plant growth hormone that is most abundant during germination and seed maturation. In flowering years when ripening seeds are available, increased gibberellic acid ingestion could amplify the effects of improved food quality on population growth by stimulating breeding before winter. Germination of that same seed crop under snow, which is known to be possible for a number of plant species,[16] might also stimulate breeding during winter in the subnivean environment, though insufficient data presently exist on the periodicity of winter breeding to link it conclusively with plant growth cycles. Nevertheless, winter breeding of the root vole in northern Finland has been known to occur only when young shoots of cotton grass were found just prior to winter with leaves newly opened.[17]

Finally under the heading of plant-animal interactions, we come to a problem of considerable controversy—carbon dioxide buildup under snow and its possible effects on subnivean animal activity.

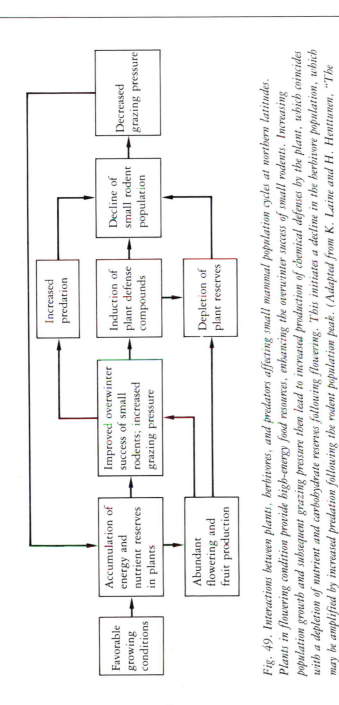

Fig. 49. Interactions between plants, herbivores, and predators affecting small mammal population cycles at northern latitudes. Plants in flowering condition provide high-energy food resources, enhancing the overwinter success of small rodents. Increasing population growth and subsequent grazing pressure then lead to increased production of chemical defenses by the plant, which coincides with a depletion of nutrient and carbohydrate reserves following flowering. This initiates a decline in the herbivore population, which may be amplified by increased predation following the rodent population peak. (Adapted from K. Laine and H. Henttunen, "The Role of Plant Production in Microtine Cycles in Northern Fennoscandia," Oikos, 40 (1983): 407–18.)

The respiration of plants buried under snow (including the root systems of large trees as well as the small perennial herbs of the ground layer), coupled with the activity of decay organisms that break down the dead organic material of the forest floor, results in the liberation of CO_2 under the snowpack. In some cases, the release of CO_2 during winter has been found to be negligible,[18] but in other cases it may be considerable.[19] If the snowpack were to act as a barrier to the diffusion of CO_2 in those instances where its evolution is substantial, then it might be expected that animals active in the subnivean environment would suffer some deleterious effects from CO_2 buildup.

In 1946 the Soviet ecologist A. N. Formozov wrote of vertical tunnels constructed in the snowpack by voles that "probably serve to ventilate the deeper parts of the burrow."[20] He also observed that "after each fresh snowfall the voles clear these 'windows,' but very rarely come out of them." The notion that subnivean mammals deliberately construct ventilation shafts to diffuse CO_2 accumulating under the snowpack has persisted in the folklore of winter ecology since that time. But some of the observations relating to the use of these tunnels are contradictory and the few attempts at measuring subnivean CO_2 levels during winter have produced inconsistent results. By one account, ventilation shafts are used only in the early part of winter and become uncommon after mid winter.[21] If their use is related to CO_2 accumulation, the reasons for their disappearance after mid winter is perplexing. Yet in a separate study at Barrow, Alaska, subnivean CO_2 levels were found to climb steadily after the first snowfall, to a maximum of 30 ppm above ambient in December, and then drop sharply to just 8 ppm above ambient, where it remained for the next four months.[22] At the Kilpisjarvi Biological Station in Finland, K. Korhonen counted anywhere from 8 to 18 "air vents" per 100 m^2 of snow surface—an unusually high number—but never found CO_2 levels under the snow high enough to have any physiological effect on the voles.[23] He noted, too (contrary to Formozov's observation), that voles showed no haste in reopening holes that had been filled with snow, and that many areas that supported large vole populations had no vents at all.

Korhonen also reported that CO_2 diffused very readily in snow, even under wet or highly compacted snow. Workers at the University of Manitoba in Canada, however, measured consistent increases in CO_2 in some habitats and reported animal movements away from those areas in which a buildup was seen.[24] So the questions remain: Do CO_2 levels under the snow at least occasionally reach deleterious levels and, if so, under what conditions (of weather, habitat, snowcover, plant activity)? How do subnivean mammals respond physiologically or behaviorally to such stresses if they do occur? And what role might this play in the overwintering success of these animals?

7 HUMANS IN COLD PLACES

We have now touched upon the majority of stresses associated with winter and have examined the varied ways in which winter-active organisms deal with the rigors of the season. We have seen many evolutionary options exploited, many different answers to a common set of environmental constraints. Our survey is almost complete. Now it seems appropriate to ask, "what of ourselves?" Humans, warm-blooded homeotherms, conspicuous denizens of snow country— what adaptations do we possess for dealing with the winter environment? How well suited physiologically are we to the cold places that we inhabit?

Now and again stories are recorded of humans in cold places showing remarkable tolerance to low temperature: Australian aborigines sleeping comfortably on the ground, nearly naked, at temperatures close to freezing; Nepalese and Andean mountain people in cold weather wearing only thin clothing without gloves or footwear; Korean women, the Ama, who dive in cold water year-round; and Inuit and other northern inhabitants who work with their bare hands at temperatures that would leave most people's fingers numb and useless. These reports notwithstanding, humans for the most part show surprisingly little physiological adaptation to low temperature. The apparent tolerance to cold in many of these northern and mountain peoples is still largely a matter of behavioral or cultural adaptation. Clothing and shelter remain our first line of defense against the cold and these we utilize with unmatched ingenuity. But take away the covers and we display only modest abilities to control the production and loss of body heat.

The most effective physiological mechanism that we have for reducing heat loss is the constriction of blood vessels situated close to the skin surface. Vasoconstriction is mediated by sympathetic nerve fibers that release noradrenaline into the vessels in response to signals from cold receptors in the skin (fig. 50). Constriction of surface vessels then shunts blood back through deeper veins and reduces heat loss as a result of decreased radiation and conduction from a cooler skin surface. When danger of freezing is imminent, extreme vasoconstriction is followed by short intervals of warming due to

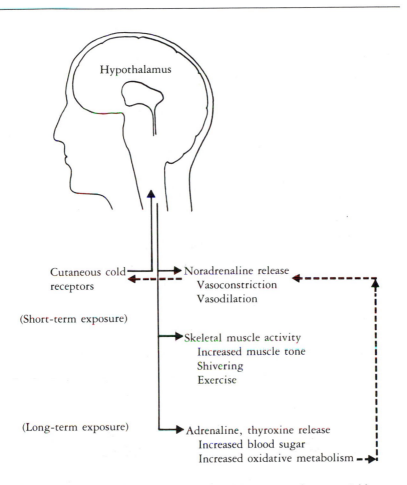

Fig. 50. Short- and long-term responses to cold exposure in humans. Cold receptors in the skin transmit nerve signals to the hypothalamus, our thermoregulatory control center, which then mediates the responses indicated. Increased heat production feeds back through skin receptors (dotted line) to regulate the input signal. (Redrawn from A. Holdcroft, Body Temperature Control in Anaesthesia, Surgery and Intensive Care *(London: Bailliere Tindall, 1980).)*

intermittent dilation of vessels. In the skin of the fingers and toes are vascular channels called "arteriovenous anastomoses" that connect directly between the arterioles and venules and provide for a bypassing of the capillary network that lies just under the skin surface. These channels are surrounded by circular smooth muscle with numerous nerve endings under involuntary control. When an anastomosis dilates to permit flow, the arteriolar muscle immediately beyond it contracts to shunt blood through the channel, short-circuiting the capillaries and flooding the extremity with warm blood.[1]

This works somewhat like the countercurrent mechanism discussed earlier and, in fact, some authors talk of countercurrent heat exchange in humans. But histological evidence for the presence of a well-developed rete mirable such as found in the appendages of many animals is lacking. Though a temperature gradient on the order of 10° from core to extremities is often observed in humans, this is more likely the result of progressive heat loss to the outside by conduction along the length of the limb. In any case, the reduction in temperature of the extremities seen in humans does not approach the very steep temperature gradient seen in the appendages of animals with well-developed countercurrent systems.

Curiously, in some northern peoples it seems that vasodilation in the extremities, rather than constriction of surface vessels, is the normal response to cold. The fingers of Inuit adults, when immersed in ice water, warm after only brief exposure and may be maintained 4° to 5° higher than the fingers of other subjects unaccustomed to the cold.[2] The increase in blood flow with vasodilation (which can reach as high as one hundred times the minimum flow to superficial tissues!)[3] greatly accelerates heat loss, but has the adaptive advantage of increasing manual dexterity at low temperatures. Apparently the increased heat loss with vasodilation can be afforded in the case of the Inuit when other mechanisms (i.e., cultural) for conserving heat are so effective. There is some reason to believe, too, that the increased flow of blood in the hands of these and other people such as Norwegian fishermen may be conditioned by physical exercise, especially at low temperature, since the same can be influenced by fitness training.[4]

On the heat production side of the equation, we fare a little better, but still lack some of the regulatory mechanisms exhibited by other mammals of cold winter climates. Increased muscle activity, either voluntarily or through shivering, remains our principal mechanism for increasing heat production. The skeletal muscle that makes up roughly 50% of our body mass contributes about 20% of our total heat production while relaxed, but vigorous exercise can increase this heat production as much as tenfold. Shivering by itself can raise the metabolic rate fivefold. This response, as with vasodilation, is also triggered by cold receptors in the skin that transmit nerve signals to the hypothalamus, our thermoregulatory control center (fig. 50). Maximum shivering response seems to occur at a skin temperature of 20° C.[5] But even before shivering, a measurable increase in muscle tone occurs, often felt as a tightening of the neck and shoulder muscles, and this alone can double heat production.

With extended exposure to cold, a nonshivering response may be elicited in humans wherein basal heat production rises in relation to increased glandular activity. Experiments with human subjects indicate that hormones secreted by the thyroid gland (e.g., thyroxin) and by the pituitary and adrenal glands (e.g., adrenaline and noradrenaline) are somehow responsible for a prolonged increase in metabolic rate over time.[6] Recall that the nonshivering response seen earlier in small mammals was also mediated by the release of noradrenaline, which triggered an increase in heat production through brown fat metabolism. However, reports of brown fat in adults are rare, though it has been found in the area of the kidneys. While it is tempting to suggest that a reversion of white fat to brown might occur in adults under cold stress, as has been seen in laboratory rats, evidence is so far lacking. Only in infants, where brown fat may comprise 1 to 6% of their body weight, is this mechanism for nonshivering thermogenesis significant in the maintenance of body temperature.[7] The most likely explanation for elevated heat production in adults is that cold stimulation of hormonal activity stimulates the conversion of glycogen to glucose (mediated by adrenaline) and increases the rate of oxidative metabolism (mediated by thyroxine), resulting in increased heat output by the liver.[8] It should be emphasized, however, that nonshivering thermogenesis in humans is of minor importance relative to the role of muscle activity in increasing heat production under cold stress.

With long-term acclimatization humans are able to increase their tolerance of the cold. There are numerous reports of lowered shivering thresholds in human subjects conditioned by prolonged exposure, one of the more interesting being of white students in Fairbanks, Alaska, who, for philosophical reasons, customarily went barefoot and lightly clothed in winter without experiencing great discomfort.[9] But as Lawrence Irving, who studied these and numerous other human subjects in the North, has noted, it is only under the most pressing of circumstances that the human species accepts the degree of prolonged exposure necessary to successfully acclimatize to the cold. Instead, most individuals go to great lengths to avoid such exposure. As often expressed by the Inuit, the best way in the end to deal with cold weather is to "take care not to be cold."[10] By our cultural and technological ingenuity, we have exploited the coldest places on earth. Biologically, however, we remain essentially tropical beings.

NOTES

Chapter 1. Winter Paths

1. K. Schmidt-Nielsen, *Animal Physiology: Adaptation and Environment*, 3rd ed. (New York: Cambridge University Press, 1983), 181.
2. Ibid., 215.
3. M. Aleksiuk, "Manitoba's Fantastic Snake Pits," *National Geographic* vol. 148, no. 5 (1975): 715–23.
4. W. D. Schmid, "Survival of Frogs in Low Temperature," *Science* 215 (1982): 697–98.

Chapter 2. The Changing Snowpack

1. P. J. Marchand, "Light Extinction under a Changing Snowcover," in J. F. Merritt, ed., *Winter Ecology of Small Mammals*, Carnegie Museum of Natural History, Spec. Pub. 10 (Pittsburgh, 1984), 33–37.
2. C. W. Thomas, "On the Transfer of Visible Radiation through Sea Ice and Snow," *Journal of Glaciology* 4 (1963): 481–84; and H. Curl, J. T. Hardy, and R. Ellermeier, "Spectral Absorption of Solar Radiation in Alpine Snowfields," *Ecology* 53 (1972): 1189–94.
3. Curl et al., "Spectral Absorption," 1189–94.
4. S. G. Richardson, and F. B. Salisbury, "Plant Responses to the Light Penetrating Snow," *Ecology* 58 (1977): 1152–58.
5. It should be noted in passing that one frequently cited report in the literature of small mammal ecology contains conflicting evidence: L. N. Evernden, and W. A. Fuller, "Light Alteration Caused by Snow and Its Importance to Subnivean Rodents," *Canadian Journal of Zoology* 50 (1972): 1023–32. To measure light quality under snow, Evernden and Fuller employed a color temperature device in which the sensor temperature varied as a function of the wavelength of light absorbed. Using this technique, they observed an apparent shift in light penetration toward the red end of the spectrum under snow of increasing depth. The authors also studied the response of captive voles in a laboratory to light of different color and found that red light inhibited sexual maturation whereas blue or white light stimulated it. Putting the two results together suggested a mechanism for regulating repro-

ductive activity under the snow. Unfortunately, the color temperature device appears to have been responding to something other than light (probably the snowpack temperature gradient). Furthermore, the authors indicate that the voles under the red light regime in the laboratory were possibly responding to a perceived total darkness rather than to color per se, since they appeared blind to the researcher's hand.

Chapter 3. Plants and the Winter Environment

1. J. R. Havis, "Water Movement in Woody Stems during Freezing, *Cryobiology* 8 (1971): 581–85.
2. P. L. Steponkus, and S. C. Wiest, "Plasma Membrane Alterations Following Cold-Acclimation and Freezing," in P. H. Li and A. Sakai, eds., *Plant Cold Hardiness and Freezing Stress: Mechanisms and Crop Implications* (New York: Academic Press, 1978).
3. J. P. Palta, and P. H. Li, "Cell Membrane Properties in Relation to Freezing Injury," in P. H. Li and A. Sakai, eds., *Plant Cold Hardiness and Freezing Stress: Mechanisms and Crop Implications* (New York: Academic Press, 1978).
4. P. J. Quin, "A Lipid-Phase Separation Model of Low-Temperature Damage to Biological Membranes," *Cryobiology* 22 (1985): 128–46.
5. C. J. Weiser, "Cold Resistance and Injury in Woody Plants," *Science* (1970): 1269–78.
6. J. M. Schmitt, M. J. Schramm, H. Pfanz, S. Coughlan, and U. Heber, "Damage to Chloroplast Membranes during Dehydration and Freezing," *Cryobiology* 22 (1985): 93–104.
7. A. G. Hirsh, R. J. Williams, and H. T. Meryman, "A Novel Method of Natural Cryoprotection," *Plant Physiology* 79 (1985): 41–56.
8. W. C. White and C. J. Weiser, "The Relation of Tissue Desiccation, Extreme Cold, and Rapid Temperature Fluctuations to Winter Injury of American Arborvitae, "*American Society of Horticultural Science* 85 (1964): 554–63.
9. P. Wardle, "Winter Desiccation of Conifer Needles Simulated by Artificial Freezing," *Arctic and Alpine Research* 13 (1981): 419–23.
10. V. C. LaMarche and H. A. Mooney, "Recent Climatic Change and Development of the Bristlecone Pine (*Pinus longaeva* Bailey) Krummholz Zone, Mt. Washington, Nevada," *Arctic and Alpine Research* 4 (1972): 61–72.
11. A. Kacperska-Palacz, "Mechanism of Cold Acclimation in Herbaceous Plants," in P. H. Li and A. Sakai, eds., *Plant Cold Hardiness and Freez-*

ing Stress: Mechanisms and Crop Implications (New York: Academic Press, 1978).

12. Weiser, "Cold Resistance," 1269–78.

13. Ibid.

14. F. B. Salisbury and C. W. Ross, *Plant Physiology* (Belmont, Ca.: Wadsworth Publishing Co., 1985).

15. Weiser, "Cold Resistance," 1269–78.

16. U. Heber, "Freezing Injury in Relation to Loss of Enzyme Activities and Protection against Freezing," *Cryobiology* 5 (1968): 188–201.

17. M. Senser, "Frost Resistance in Spruce (*Picea abies* (L.) Karst): III. Seasonal Changes in the Phospho- and Galacto-Lipids of Spruce Needles," *Zeitschrift für Pflanzenphysiologie* 105 (1982): 229–39.

18. Ibid.

19. Kacperska-Palacz, "Mechanism of Cold Acclimation."

20. W. Larcher and H. Bauer, "Ecological Significance of Resistance to Low Temperature," in O. L. Lange, P. S. Nobel, C. B. Osmond, and H. Ziegler, eds., *Physiological Plant Ecology I*. Encyclopedia of Plant Physiology, New Series, vol. 12a (Berlin: Springer-Verlag, 1981), 403–37.

21. Weiser, "Cold Resistance," 1269–78.

22. Larcher and Bauer, "Ecological Significance," 403–37.

23. S. Eiga and A. Sakai, "Altitudinal Variation in Freezing Resistance of Saghalien Fir (*Abies sachalinensis*)," *Canadian Journal of Botany* 62 (1984): 156–60.

24. A. F. W. Schimper, *Plant-Geography upon a Physiological Basis*. trans. W. R. Risher (Oxford: Clarendon, 1903).

25. W. Tranquillini, *Physiological Ecology of the Alpine Timberline* (Berlin: Springer-Verlag, 1979).

26. A. Sakai, "Mechanism of Desiccation Damage of Conifers Wintering in Soil Frozen Areas," *Ecology* 51 (1970): 657–64.

27. P. S. Nobel, *Biophysical Plant Physiology and Ecology* (San Francisco: W. H. Freeman and Co., 1983).

28. P. Holmgren, P. G. Jarvis, and M. S. Jarvis, "Resistances to Carbon Dioxide and Water Vapor Transfer in Leaves of Different Plant Species," *Physiologia Plantarum* 18 (1965): 557–73.

29. Nobel, *Biophysical Plant Physiology*.

30. P. J. Marchand and B. F. Chabot, "Winter Water Relations of Treeline Plant Species on Mt. Washington, New Hampshire," *Arctic and Alpine Research* 10 (1978): 105–16.

31. Ibid., 105–16.

32. D. T. Kincaid and E. E. Lyons, "Winter Water Relations of Red Spruce on Mount Monadnock, New Hampshire," *Ecology* 62 (1981): 1155–61.

33. Marchand and Chabot, "Winter Water Relations," 105–16.

34. H. T. Hammel, "Freezing of Xylem Sap without Cavitation," *Plant Physiology* 42 (1967): 55–66.

35. S. C. Gregory and J. A. Petty, "Valve Action of Bordered Pits in Conifers," *Journal of Experimental Botany* 24 (1973): 763–67.

36. Havis, "Water Movement in Woody Stems," 581–85.

37. G. Hygen, "Water Stress in Conifers during Winter," in B. Slavik, ed., *Water Stress in Plants* (The Hague: W. Junk, 1963).

38. N. Polunin, "Conduction through Roots in Frozen Soil," *Nature* 132 (1933): 313–14.

39. P. W. Owston, J. L. Smith, and H. G. Halverson, "Seasonal Water Movement in Tree Stems," *Forest Science* 18 (1972): 266–72.

40. Nobel, *Biophysical Plant Physiology* 74 and 483.

41. Ibid., 514.

42. Marchand and Chabot, "Winter Water Relations," 105–16.

43. U. Benecke and W. M. Havranek, "Gas-Exchange of Trees at Altitudes Up to Timberline, Craigieburn Range, New Zealand," in U. Benecke and M. R. Davis, eds., *Mountain Environments and Subalpine Tree Growth,* New Zealand Forest Service Technical Paper No. 70 (1980), 195–212.

44. Tranquillini, *Physiological Ecology.*

45. E. D. Schulze, H. A. Mooney, and E. L. Dunn, "Wintertime Photosynthesis of Bristlecone Pine (*Pinus aristata*) in the White Mountains of California," *Ecology* 48 (1967): 1044–47.

46. J. F. Chabot and B. F. Chabot, "Developmental and Seasonal Patterns of Mesophyll Ultrastructure in *Abies balsamea,*" *Canadian Journal of Botany* 53 (1975): 295–304.

47. Marchand and Chabot, "Winter Water Relations," 105–16.

48. L. L. Tieszen, "Photosynthetic Competence of the Subnivean Vegetation of an Arctic Tundra," *Arctic and Alpine Research* 6 (1974): 253–56.

49. E. Mäenpää, "Photosynthesis and Respiration of Bilberry (*Vaccinium myrtillus* L.) under Winter Conditions," unpublished manuscript.

50. Tranquillini, *Physiological Ecology.*

51. T. O. Perry, "Winter-Season Photosynthesis and Respiration by Twigs and Seedlings of Deciduous and Evergreen Trees," *Forest Science* 17 (1971): 41–43.

52. Ibid.
53. M. Schaedle and K. C. Foote, "Seasonal Changes in the Photosynthetic Capacity of *Populus tremuloides* Bark," *Forest Science* 17 (1971): 308–13.
54. S. Eurola and A. Huttunen, "Riisitunturi," *Oulanka Reports* 5 (1984): 65–67.
55. J. L. Hadley and W. K. Smith, "Wind Effects on Needles of Timberline Conifers: Seasonal Influence on Mortality," *Ecology* 67 (1986): 12–19.
56. P. J. Marchand, F. L. Goulet, and T. C. Harrington, "Death by Attrition: An Hypothesis for Wave-Mortality of Subalpine *Abies balsamea*," *Canadian Journal of Forest Research* 16 (1986): 591–96.

Chapter 4. Animals and the Winter Environment

1. D. D. Feist, "Metabolic and Thermogenic Adjustments in Acclimatization of Subarctic Alaskan Red-backed Voles," in J. F. Merritt, ed., *Winter Ecology of Small Mammals,* Carnegie Museum of Natural History, Spec. Pub. 10 (Pittsburgh, 1984), 131–38.
2. J. O. Wolf and W. Z. Lidicker, Jr., "Communal Winter Nesting and Food Sharing in Taiga Voles," *Behavioral Ecology and Sociobiology* 9 (1981): 237–40.
3. D. M. Madison, "Group Nesting and Its Ecological and Evolutionary Significance in Overwintering Microtine Rodents," in J. F. Merritt, ed., *Winter Ecology of Small Mammals,* Carnegie Museum of Natural History, Spec. Pub. 10 (Pittsburgh, 1984), 267–74.
4. Madison, "Group Nesting," 267–74.
5. L. Irving, *Arctic Life of Birds and Mammals, Including Man* (Berlin: Springer-Verlag, 1972).
6. Ibid.
7. J. S. Hart and O. Heroux, "A Comparison of Some Seasonal and Temperature-Induced Changes in *Peromyscus*: Cold Resistance, Metabolism, and Pelage Insulation," *Canadian Journal of Zoology* 31 (1953): 528–34.
8. Irving, *Arctic Life.*
9. B. A. Wunder, "Strategies for, and Environmental Cueing Mechanisms of, Seasonal Changes in Thermoregulatory Parameters of Small Mammals," in J. F. Merritt, ed., *Winter Ecology of Small Mammals,* Carnegie Museum of Natural History, Spec. Pub. 10 (Pittsburgh, 1984), 165–72.
10. Irving, *Arctic Life.*

11. S. A. Richards, *Temperature Regulation* (London: Wykeham Publications, 1973).

12. Schmidt-Nielsen, *Animal Physiology,* 291.

13. Feist, "Metabolic and Thermogenic Adjustments," 131–38.

14. Wunder, "Strategies for, and Environmental Cueing Mechanisms of, Seasonal Changes" 165–72; and J. F. Merritt, "Winter Survival Adaptations of the Short-tailed Shrew (*Blarina brevicauda*) in an Appalachian Montane Forest," *Journal of Mammalogy* 67 (1986): 450–64.

15. Feist, Metabolic and Thermogenic Adjustments," 131–38.

16. V. Nolan, Jr., and E. D. Ketterson, "An Anaylsis of Body Mass, Wing Length, and Visible Fat Deposits of Dark-eyed Juncos Wintering at Different Latitudes," *Wilson Bulletin* 95 (1983): 603–20.

17. C. Carey, W. R. Dawson, L. C. Maxwell, and J. A. Faulkner, "Seasonal Acclimatization to Temperature in Cardueline Finches. II. Changes in Body Composition and Mass in Relation to Season and Acute Cold Stress," *Journal of Comparative Physiology* 125 (1978): 101–13.

18. Nolan and Ketterson, "An Analysis of Body Mass," 603–20.

19. W. R. Dawson and C. Carey, "Seasonal Acclimatization to Temperature in Cardueline Finches. I. Insulative and Metabolic Adjustments," *Journal of Comparative Physiology* 112 (1976): 317–33.

20. G. C. West and J. S. Hart, "Metabolic Responses of Evening Grosbeaks to Constant and to Fluctuating Temperatures," *Physiological Zoology* 39 (1966): 171–84.

21. G. C. West, "Shivering and Heat Production in Wild Birds," *Physiological Zoology* 38 (1965): 111–20.

22. K. Korhonen, "Microclimate in the Snow Burrows of Willow Grouse (*Lagopus lagopus*)," *Ann. Zool. Fennici* 17 (1980): 5–9.

23. W. A. Buttemer, "Energy Relations of Winter Roost-Site Utilization by American Goldfinches (*Carduelis tristis*)," *Oecologia* 68 (1985): 126–32.

24. S. B. Chaplin, "The Physiology of Hypothermia in the Black-capped Chickadee, *Parus atricapillus*," *Journal of Comparative Physiology* 112 (1976): 335–44; and West, "Shivering and Heat Production," 111–20.

25. M. W. Schwan and D. D. Williams, "Temperature Regulation in the Common Raven of Interior Alaska," *Comparative Biochemical Physiology* 60a (1978): 31–36; and West, "Shivering and Heat Production," 111–20.

26. C. Carey and R. L. Marsh, "Shivering Finches," *Natural History* 90 (1981): 58–63; and Carey et al., "Seasonal Acclimatization."

27. Carey and Marsh, "Shivering Finches," 58–63.

28. Chaplin, "Physiology of Hypothermia," 335–44.

29. Wunder, "Strategies for, and Environmental Cueing Mechanisms of, Seasonal Changes," 165–72.

30. Feist, "Metabolic and Thermogenic Adjustments," 131–38.

31. R. W. Coles, "Thermoregulatory Function of the Beaver Tail," *American Zoologist* 9 (1969): 203a.

32. D. L. Kilgore, Jr., and K. Schmidt-Nielsen, "Heat Loss from Ducks' Feet Immersed in Cold Water," *Condor* 77 (1975): 475–78.

33. J. H. Brown and R. C. Lasiewski, "Metabolism of Weasels: The Cost of Being Long and Thin," *Ecology* 53 (1972): 939–43.

34. D. C. Ure, "Autumn Mass Dynamics of Red-backed Voles (*Clethrionomys gapperi*) in Colorado in Relation to Photoperiod Cues and Temperature," in J. F. Merritt, ed., *Winter Ecology of Small Mammals,* Carnegie Museum of Natural History, Spec. Pub. 10 (Pittsburgh: 1984), 193–200.

35. B. Wunder, personal communication.

36. J. Zejda, personal communication.

37. B. K. McNab, "On the Ecological Significance of Bergmann's Rule," *Ecology* 52 (1971): 845–54.

38. Ibid.

39. W. J. Hamilton, III, and F. Heppner, "Radiant Solar Energy and the Function of Black Homeotherm Pigmentation: An Hypothesis," *Science* 155 (1967): 196–97.

40. W. J. Hamilton, III, *Life's Color Code* (New York: McGraw-Hill Book Co., 1973), 29–83.

41. Irving, *Arctic Life.*

Chapter 5. Life Under Ice

1. K. Schmidt-Nielsen, *Animal Physiology,* 245.

2. Ibid.

3. R. E. Gatten, Jr., "Anaerobiosis in Amphibians and Reptiles," *American Zoologist* 25 (1985): 945–54.

4. K. Schmidt-Nielsen, *Animal Physiology,* 184–86.

5. R. T. Wright, "Dynamics of a Phytoplankton Community in an Ice-covered Lake," *Limnology and Oceanography* 9 (1964): 163–78.

6. C. W. Boylen and R. B. Sheldon, "Submergent Macrophytes: Growth under Winter Ice Cover," *Science* 194 (1976): 841–42.

7. P. J. Marchand, "Oxygen Evolution by *Elodea canadensis* under Snow and Ice Cover: A Case for Winter Photosynthesis in Subnivean Vascular Plants," *Aquilo Ser. Botanica* 23 (1985): 57–61.

Chapter 6. Plant-Animal Interaction

1. K. Danell, K. Huss-Danell, and R. Bergstrom, "Interactions between Browsing Moose and Two Species of Birch in Sweden," *Ecology* 66 (1985): 1867–78.

2. J. P. Bryant, G. D. Wieland, T. Clausen, and P. Kuropat, "Interactions of Snowshoe Hare and Feltleaf Willow in Alaska," *Ecology* 66 (1985): 1564–73.

3. J. P. Bryant, F. S. Chapin, III, and D. R. Klein, "Carbon/Nutrient Balance of Boreal Plants in Relation to Vertebrate Herbivory," *Oikos* 40 (1983): 357–68.

4. P. B. Reichardt, J. P. Bryant, T. P. Clausen, and G. D. Wieland, "Defense of Winter-Dormant Alaska Paper Birch against Snowshoe Hares," *Oecologia* 65 (1984): 58–69.

5. Ibid.

6. Bryant et al., "Interactions of Snowshoe Hare," 1564–73.

7. J. Tahvanainen, E. Helle, R. Julkunen-Tiitto, and A. Lavola, "Phenolic Compounds of Willow Bark as Deterrents against Feeding by Mountain Hare," *Oecologia* 65 (1985): 319–23.

8. Tahvanainen et al., "Phenolic Compounds," 319–23.

9. L. B. Keith, "Role of Food in Hare Population Cycles," *Oikos* 40 (1983): 385–95; and Bryant et al., "Carbon/Nutrient Balance," 357–68.

10. K. Laine, and H. Henttunen, "The Role of Plant Production in Microtine Cycles in Northern Fennoscandia," *Oikos* 40 (1983): 407–18.

11. S. Eurola, H. Kyllonen, and K. Laine, "Plant Production and Its Relation to Climatic Conditions and Small Rodent Density in Kilpisjarvi Region (69°05′N, 20°40′E), Finnish Lapland," in J. F. Merritt, ed., *Winter Ecology of Small Mammals,* Carnegie Museum of Natural History, Spec. Pub. 10 (Pittsburgh, 1984), 121–30.

12. Laine and Henttunen, "The Role of Plant Production," 407–18.

13. Ibid.

14. Keith, "Role of Food," 385–95.

15. P. Berger, N. C. Negus, E. H. Sanders, and P. D. Gardner, "Chemical Triggering of Reproduction in *Microtus mountanus,*" *Science* 214 (1981):

69–70; and P. Olsen, "The Stimulating Effect of a Phytohormone, Gibberellic Acid, on Reproduction of *Mus musculus*," *Australian Wildlife Research* 8 (1981): 321–15.

16. F. B. Salisbury, "Light Conditions and Plant Growth under Snow," in J. F. Merritt, ed., *Winter Ecology of Small Mammals*, Carnegie Museum of Natural History, Spec. Pub. 10 (Pittsburgh: 1984), 39–50.

17. J. Tast and A. Kaikusalo, "Winter Breeding of the Root Vole, *Microtus oeconomus*, in 1972/1973 at Kilpisjarvi, Finnish Lapland," *Ann. Zool. Fennici* 13 (1976): 174–78.

18. W. A. Reiners, "CO_2 Evolution from the Floor of Three Minnesota Forests," *Ecology* 49 (1968):471–83.

19. P. Havas and E. Mäenpää, "Evolution of Carbon Dioxide at the Floor of a *Hylocomium-Myrtillus* Type Spruce Forest," *Aquilo Ser. Botanica* 11 (1972): 4–22.

20. A. N. Formozov, *Snow Cover as an Integral Factor of the Environment and its Importance in the Ecology of Mammals and Birds*, trans. W. Prychodko and W. O. Pruitt, Jr., Boreal Institute, Occasional Pub. No. 1 (Edmonton: Univ. of Alberta, 1946).

21. Evernden and Fuller, "Light Alteration Caused by Snow," 1023–32.

22. J. J. Kelley, Jr., D. F. Weaver, and B. P. Smith, "The Variation of Carbon Dioxide under the Snow in the Arctic," *Ecology* 49 (1968): 358–61.

23. K. Korhonen, "Ventilation in the Subnivean Tunnels of the Voles *Microtus agrestis* and *M. oeconomus*," *Ann. Zool. Fennici* 17 (1980): 1–4.

24. C. E. Penny and W. O. Pruitt, Jr., "Subnivean Accumulation of CO_2 and Its Effects on Winter Distribution of Small Mammals," in J. F. Merritt, ed., *Winter Ecology of Small Mammals*. Carnegie Museum of Natural History, Spec. Pub. 10 (Pittsburgh: 1984), 373–80.

Chapter 7. Humans in Cold Places

1. S. A. Richards, *Temperature Regulation* Wykeham (London: Wykeham Publications Ltd., 1973), 77.

2. C. J. Eagan, "Introduction and Terminology: Habituation and Peripheral Tissue Adaptations," in E. F. Moran, *Human Adaptability* (Boulder Co.: Westview Press, 1982), 121.

3. A. Holdcroft, *Body Temperature Control in Anaesthesia. Surgery and Intensive Care* (London: Bailliere Tindall, 1980).

4. O. G. Edholm and A. L. Bacharach, *The Physiology of Human Survival* (London: Academic Press, 1965).

5. Holdcroft, *Body Temperature Control*.

6. C. C. VanWie, "Physiological Responses to Cold Environments," in J. D. Ives and R. G. Barry, eds., *Arctic and Alpine Environments,* (London: Methuen, 1974).
7. Holdcroft, *Body Temperature Control.*
8. Ibid.
9. Irving, "Arctic Life."
10. Ibid.

GLOSSARY

Acclimation. Physiological adjustment to changing environmental conditions, especially to increasing or decreasing temperature. This term is used in the plant literature in reference to seasonal changes in cold tolerance, but animal researchers restrict its use to shorter-term adjustments, usually observed under laboratory conditions (*see* Acclimatization).

Acclimatization. Seasonal or long-term physiological adjustment, usually in response to temperature changes. Preferred in this context by animal researchers over the term "acclimation."

Ambient. Referring to near surroundings. Characteristic of the immediate environment in which a plant or animal operates.

Basal metabolic rate (BMR). The lowest rate of metabolism shown by a homeotherm while fasting and at rest within its thermal neutral zone.

Boundary layer. Thin shell of air surrounding an object and having properties (e.g., temperature, humidity) intermediate between the surface of the object and the bulk atmosphere around it.

Brown fat. Adipose tissue characterized by a yellow to light brown color, having a high concentration of mitochondria and hence capable of high oxidation rates and heat production.

Cavitation. Formation of a gas-filled bubble or cavity in a liquid, often occurring in the narrow capillaries of a plant as a result of the dissolution of gasses upon freezing.

Conduction. Transfer of heat via molecular collisions.

Constructive metamorphism. The process in which ice crystals favorably situated in the snowpack grow by accretion of water onto their surfaces.

Convection. Transfer of heat via a moving fluid.

Cuticle. A layer of wax covering the outer surface of a plant leaf, serving primarily to reduce water loss.

Denaturation. Alteration or loss of normal biotic function.

Depth Hoar. Brittle ice crystals, often hollow and cuplike, formed in warmer layers of the snowpack as a result of continuous vapor loss from their surfaces.

Desiccation. Extreme and damaging water loss.

Destructive metamorphism. The process in which new-fallen snow crystals lose

167

their delicate structure by a redistribution of internal energy, and co-alesce into rounded ice grains.

Diffuse porous. A characteristic ring pattern in the wood (xylem tissue) of some broadleaf tree species in which individual conducting cells are of small size and vary little in diameter from early spring to late summer, thus showing only a relatively faint line of demarcation between one year's growth and the next (cf. Ring porous).

Exothermic. A reaction or change of state in which heat is given off.

Extracellular. Outside of the cell (synonomous with Intercellular).

Glycolosis. The breakdown of carbohydrate in the absence of oxygen to produce ATP and lactic acid.

Herbaceous. Nonwoody.

Hibernaculum. Any refuge used over winter by a hibernating animal.

Homeotherm. An animal that maintains a body temperature within a high and relatively narrow range independently of the temperature of its surroundings. Often referred to as a "warm-blooded" animal.

Hypothermia. A condition in which the body core temperature falls below that considered normal for a homeothermic animal.

Insolation. Incoming solar radiation.

Isozyme. Multiple forms of the same enzyme, usually having different temperature optima and, thus, varying in their ability to catalyze a reaction at a given temperature.

Intercellular. Between cells (synonomous with Extracellular).

Intracellular. Within cells.

Latent heat. A quantity of heat tied up or released with phase changes of a substance, but not affecting the temperature of the substance.

Lipids. Any of numerous fats or fatlike substances consisting of water-insoluable hydrocarbon molecules that comprise, along with proteins and carbohydrate, the principal structural material of living cells.

Littoral zone. That portion of a lake forming the interface between land and deep water, defined by the shoreline at one extent and the limits of submerged, rooted aquatic plants at the other.

Long-wave energy. Radiant energy having a wavelength greater than visible light. In an ecological context usually referring to infrared or heat energy.

Lower critical temperature (LCT). That temperature at which a homeotherm can no longer maintain normal body temperature by passive means and must increase metabolic heat production.

Melanin. A granular pigment varying in color from yellow to black that is present in the feathers and hair of many animals.

Melt metamorphism. A process of change in snowpack structure characterized by melting and refreezing and accompanied by a marked increase in snowpack density.

Microclimate. The climate close to the ground or in the immediate vicinity of an organism (e.g., the nest microclimate).

Microtine. Any rodent in the subfamily *Microtinae,* characteristically having a heavyset body, short legs, and a tail usually shorter than half the head and body length.

Mitochondria. Small rod-shaped or oval organelles in the cell cytoplasm that utilize oxygen and produce most of the ATP, the energy currency of the cell.

Nonshivering thermogenesis. The capacity in homeotherms to increase heat production without muscular activity.

Noradrenaline. A hormone secreted by the adrenal medulla that mediates several thermoregulatory functions, including vasodilation, vasoconstriction, and brown fat metabolism. Also known as norepinephrine.

Overturn. An event or process marked by the reduction of temperature and density differences between the surface and bottom waters of a lake such that complete mixing of the water occurs under the influence of wind-generated currents.

Pelage. The coat (i.e., fur or hair) of a mammal.

Permafrost. A thermal condition of earth materials in which the temperature of soil, rock, or water remains belw 0° C through the entire year.

Photoperiod. Amount of light received daily. Total length of time between sunrise and sunset.

Plasma membrane. The semipermeable membrane separating the cytoplasm from the wall of a plant cell.

Poikilotherm. An organism whose body temperature fluctuates with that of its surroundings, except as it may be regulated behaviorally by such means as basking in the sun. Often referred to as a "cold-blooded" animal.

Radiation. The propagation of energy through space. Also, that energy received by or emitted from a radiating object.

Rete mirable. "Miraculous net." A dense network of veins and arteries in close contact with each other that facilitates transfer of heat from arterial blood to cold venous blood returning from the extremities, thus reducing heat loss to an animal's surroundings.

Rime ice. Ice that is formed as supercooled water droplets in clouds impact and freeze instantly on exposed objects.

Ring porous. A characteristic ring pattern in the wood (xylem tissue) of some broadleaf tree species in which individual conducting cells vary in size from the conspicuously large diameter and thin-walled cells produced in early spring to the small diameter, thick-walled cells produced in late summer, thus showing a marked line of demarcation between one year's growth and the next (cf. Diffuse porous).

Shivering thermogenesis. The capacity in homeotherms to increase heat production by involuntary muscle contractions.

Short-wave energy. Radiant energy having a wavelength within or close to the visible portion of the spectrum. Incoming solar radiation is loosely termed short-wave radiation because most of the sun's energy is emitted within this portion of the spectrum.

Sintering. The fusion of ice grains in contact with each other, tending toward the development of a solid ice mass with concomitant reduction of pore space.

Stomate. Pores on the surface of a leaf formed by a pair of specialized "guard cells" that control the size of the opening to regulate the diffusion of water vapor and carbon dioxide into and out of the leaf.

Subcutaneous. Beneath the skin.

Sublimation. A change in phase from solid directly to vapor.

Subnivean. Beneath the snowpack.

Supercooled. Having a temperature below the normal freezing point of that substance without ice formation occurring.

Temperature gradient. The difference in temperature between two points separated by a given distance. The gradient is said to be "steep" if the difference per unit distance is relatively great.

Temperature inversion. A condition in which air temperature increases with height above the ground, contrary to the normal expectation of a temperature decrease with elevation.

Thermal conductivity. The rate of heat transfer through a material of unit thickness for a given difference in temperature across the material, usually expressed as watts/cm/°K.

Thermal neutral zone (TNZ). The temperature range within which a homeotherm can maintain its basal metabolic rate through passive regulation of heat gains and losses.

Thermoregulation. Maintenance of body temperature within a preferred range by either physical or physiological means.

Torpor. A short-term condition physiologically similar to hibernation in which metabolic rate and body temperature may be reduced to conserve energy.

Vasoconstriction. Contraction of the circular muscle of arterioles, causing an increase in resistance and decrease in volume of blood flow.

Vasodilation. Opposite of vasoconstriction.

Xylem. The major conducting tissue of vascular plants, comprised mostly of interconnected, nonliving cells through which water moves passively from roots to leaves.

INDEX

12-13	DATE DUE		
12-10-07			
5-16-14			